VOLCANOES AND

A GUIDE TO OUR UNQUIET EARTH

EARTHQUAKES

**CHIARA
MARIA PETRONE**

**ROBERTO
SCANDONE**

**ALEX
WHITTAKER**

Smithsonian Books
Washington, DC

Published in Great Britain by the Natural History Museum, Cromwell Road, London SW7 5BD

Copyedited by Celia Coyne
Internal design by Mercer Design, London
Reproduction by Saxon Digital Services

Published in North America by Smithsonian Books

This book may be purchased for educational, business, or sales promotional use. For information, please write: Special Markets Department, Smithsonian Books, P.O. Box 37012, MRC 513, Washington, DC 20013

ISBN 978-1-58834-655-1

Library of Congress Cataloging-in-Publication Data

Names: Petrone, Chiara Maria, author. I Scandone, Roberto, author. I
    Whittaker, Alex, author.
Title: Volcanoes and earthquakes : a guide to our unquiet earth / Chiara
    Maria Petrone, Roberto Scandone, and Alex Whittaker.
Description: Washington, DC : Smithsonian Books, [2019] I Includes
    bibliographical references and index.
Identifiers: LCCN 2019019047 I ISBN 9781588346551 (pbk.)
Subjects: LCSH: Volcanoes--Popular works. I Earthquakes--Popular works. I
    Plate tectonics--Popular works. I Physical geography--Popular works.
Classification: LCC QE521.2 .P48 2019 I DDC 551.2--dc23 LC record available at
https://lccn.loc.gov/2019019047

Printed in China by Toppan Leefung Printing Limited, not at government expense
23 22 21 20 19   5 4 3 2 1

For permission to reproduce illustrations appearing in this book, please correspond directly with the owners of the works, as listed on page 144. Smithsonian Books does not retain reproduction rights for these images individually or maintain a file of addresses for sources.

Contents

Introduction

Volcanoes and earthquakes are amongst the most powerful natural forces and the most evident, spectacular and often highly destructive phenomena of our planet Earth. We look at volcanoes in awe and are terrified by the unpredictability and destructive power of earthquakes. Those of us that have experienced an earthquake will never forget it. I was a child when I experienced the first earthquake in my life. It turned out to be a big and a very destructive one not far from where I was living with my parents in central southern Italy. After almost forty years, I still have a very vivid memory of that scary moment, running with my family down the stairs of our four-storey building. Our dog had tried to alert us 20 minutes before the quake and then, during the event, refused to leave the flat having found a secure spot under one of the pillar arches. They were interminable seconds. Luckily, we were far from the epicentre and we were all safe with no damage to our flat or my town, but the area closer to the epicentre was devastated and the death toll was extremely high.

My first encounter with a volcano was more relaxing and magical. It happened a few years after the earthquake, when I was a teenager on a family holiday in Sicily. Lady Etna was erupting magical lava flows, so beautiful to look at that I was completely enchanted. The surrounding landscape was like walking on the moon. I was in total awe and fell in love with volcanoes immediately. When I realised that part of the sole of my beloved and totally volcano inappropriate summer shoes was slightly melted by the still-hot lava flow, I decided that I had to keep walking on the slopes of volcanoes and try to understand their secrets.

Later, I understood that it is possible to learn a lot from a single rock and even more from the minerals forming that rock. Minerals are the messengers of

OPPOSITE: Spectacular lava fountains from two different vents of the southeast crater of Etna volcano during the 12 April 2012 eruption.

the volcano, preserving a large amount of information locked in their chemical composition and textural characteristics. When we unlock their secrets, we can understand the processes and the time leading up to a volcanic eruption.

Volcanoes and earthquakes are not randomly distributed on the Earth surface, but their occurrence and location are regulated by plate tectonics, the unique engine of our planet Earth. In fact, there is no other known planet, so far, that has the plate tectonics we see on Earth. This engine has shaped the Earth in the over 4.5 billion years of its life, and is constantly changing it in a continuous cycle of creation and destruction. However, the rate of these processes is very slow and we don't have a sense that it is happening. These internal forces have made the existence of life possible, not only favouring human life but also providing us with many essential natural resources, from soil to geothermal energy to minerals.

This book aims to provide you with the key information about plate tectonics, volcanoes and earthquakes, answering some questions that you may have about these topics. Dr Alexander C. Whittaker will give you all the insights into plate tectonics, from the early, exciting days of this new theory up to the new, recent ideas that are still under development. The destructive powers of earthquakes, alongside several famous examples of well-known earthquakes, and how we can defend our lives from this immense natural power, is explored in detail by Professor Roberto Scandone. Finally, different types of volcanoes and their eruptive activity, their impact on our planet and on our lives are explored through my (Dr Chiara Maria Petrone) personal lenses without forgetting the new scientific concepts that are emerging, and how scientific knowledge can facilitate the necessary dialogue between respect for these natural phenomena and the needs of the communities living in the vicinity of an active volcano. I, together with my travel friends, Alex and Roberto, hope you will enjoy this book.

About the authors

Dr Chiara Maria Petrone is Research Leader in Petrology and Volcanology at the Natural History Museum, London. Chiara read Geology at the University of Florence, Italy where she also did a PhD on the petrology of the Mexican volcanic arc. She was a post-doctoral fellow at the University of Kyoto, Japan and at the Carnegie Institution of Washington DC, USA. Before joining the

Natural History Museum she was an electron microprobe lab manager at the CNR in Florence and at the University of Cambridge. She is currently leading the petrology and volcanology research group at Natural History Museum.

Professor Roberto Scandone is a Professor of Physical Volcanology at the University of Roma Tre. Before this, he was Professor of Geophysics at the University Federico II of Naples and a Researcher at the Vesuvius Observatory of Ercolano. Currently is an Associate Researcher at the INGV-Vesuvius Observatory.

Dr Alex Whittaker is a Senior Lecturer in Tectonics at Imperial College London. Alex read Natural Sciences at the University of Cambridge before moving to Edinburgh University to do a PhD in landscape dynamics and neo-tectonics in the Italian Apennines. Following an Entente Cordiale Fellowship at Université Joseph Fourier, France, he moved to Imperial College in 2007 where he now leads a research group in tectonics and surface processes.

Eruption of Etna volcano, Italy in 2002.

1 Tectonics

Planet Earth is a unique place. Seen from space, the distinctive arrangement of continental landmasses, separated by the inky blue expanse of ocean, is an arresting and beautiful sight. The Earth has an enormous surface area of 500 million km² (196 million sq miles), but only around 30% of that figure is land known as the 'continental crust'. The rest is ocean, covering 70% of the surface with an average depth of just under 4 km (2½ miles). Underlying this is a thin covering of deep-sea sediments and the oceanic crust. We also know that the Earth is a place of topographic extremes. This is obvious in the continents, which range from high elevation mountain belts, such as the Himalayas or the Alps, to lowland plains and forests over relatively short distances. But the same is even more true under the oceans, where depths can be relatively shallow in some places, such as the mid-Atlantic, but can plunge to more than 10 km (6 miles) in the deepest ocean trenches. In fact, from Mt Everest at nearly 9 km (5½ miles) elevation, to the famous Marianas Trench in the Pacific Ocean at approximately 11 km (7 miles) deep, the surface of the Earth varies by more than 20 km (12½ miles) in elevation. Why?

It is a fundamental question to ask why the Earth has such high mountains and deep ocean trenches, particularly as they are often found together, for example bordering the Pacific coast of South America. Even more striking is that these features are associated with active volcanoes and earthquakes. How are they linked? Has the Earth always been this way? Do these topographical and geological features evolve over short or long time periods? Have the continents and oceans always been in the same place? Today we have scientific answers to these questions, but until relatively recently these problems had no sensible explanation at all. In fact, a proper explanation was not really available until the 1960s.

OPPOSITE: Earth from space showing the continents covering 30% of the surface area and the oceans covering 70%.

The theory that emerged to explain these features, and much else about the behaviour of planet Earth as a system, is one of the most important discoveries by scientists in the twentieth century. This revolutionary theory has become known as plate tectonics. The word tectonics itself comes from the Greek work *tektonikos* (τεκτονικός), which means 'relating to building'. Tectonics had been previously used as a general term to talk about processes involved in mountain building, earthquakes and volcanism etc. Plate tectonic theory allowed these physical processes, the topography of the Earth itself, and its geological history, to be put in the context of a single unifying theory. What we now know is that the surface of the Earth is divided into at least 13 major tectonic plates and several smaller ones. These plates are generally rigid, vary in size and move at rates of several centimetres a year – roughly the pace that your fingernails grow. The movement is fast enough to be measured by satellites in space, including the kind of global positioning system (GPS) technology that is found in a smart phone or car. As we shall see, where plates move away from each other, towards each other, or past each other, mountains and oceans can be formed and earthquakes and volcanoes can occur. In this chapter we will look at where this remarkable theory came from and what it reveals about geological processes at the Earth's surface.

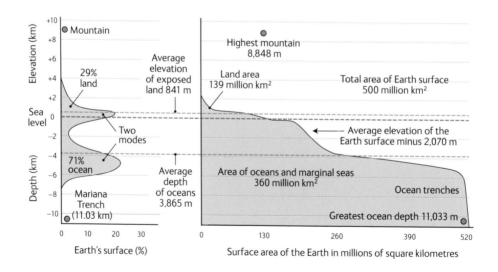

The distribution of topography above and below sea level on Earth. The graph on the left shows the percentage distribution of the Earth's surface as a function of elevation above (or depth below) sea level. The extremes are more than 20 km (12½ miles) apart.

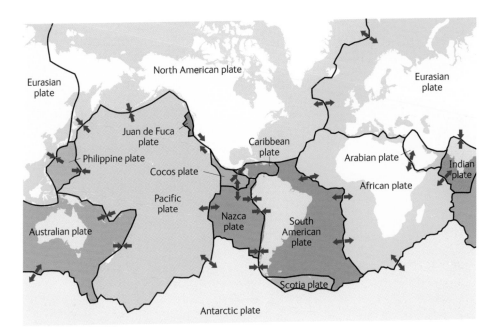

Plate tectonic map of the Earth showing the major plates.

Continental drift and the birth of an idea

The question of whether the continental landmasses and oceans are fixed in time and space has a long history. A number of natural philosophers from the sixteenth century onwards, such as Abraham Ortelius, Francis Bacon and Antonio Snider, remarked on the jigsaw-like fit of the continents on either side of the Atlantic Ocean and speculated on whether they were once joined. For instance, Ortelius in his 1596 work *Thesaurus Geographicus* proposed that North and South America were ripped away from Europe and Africa by earthquakes and floods. The Austrian geologist Eduard Suess in the late nineteenth century built on this, and argued that today's continents in the southern hemisphere once made up a single continent, which he named Gondwanaland. Perhaps the most famous of all the early proponents of continental drift is the German meteorologist and geologist Alfred Wegener, who proposed the theory (and coined the term) in a pair of research papers in 1912 and 1915. Crucially, his argument for continental drift was not just based on the similarity of the shape of the coastlines of South America and Africa. He also showed that there was a good match for geological features seen on

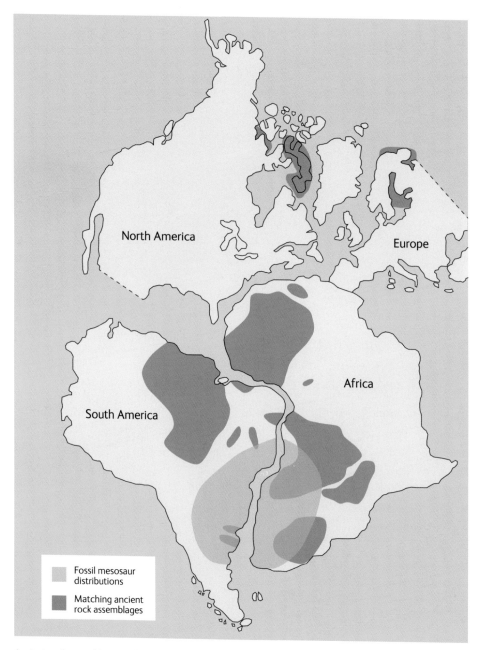

Fossil mesosaur
distributions

Matching ancient
rock assemblages

Ancient rock assemblages and distribution of fossil mesosaurs closely
match each other on either side of the Atlantic Ocean. Boundaries that are
uncertain are shown by dashed lines.

either side of the Atlantic, including crystalline and ancient rock assemblages, the distribution of glacial deposits and other trends in rock strata. Additionally, fossil evidence from palaeontology, such as the distribution of vertebrates, also supported this idea, with identical fossils suggesting an initial shared history. As the Atlantic Ocean was 'born' and the continents separated, the divergence in fossil assemblages indicated the evolution of different ecosystems over time. Much of this evidence we see as incontrovertible today.

Nevertheless, Wegener's ideas were initially resisted by many. To understand why, it is important to realize that Wegener and others did not have a plausible scientific mechanism for how continental drift might operate. Surely the Earth's crust is far too rigid to deform and drift in this way? How would it actually work? No-one had a realistic theory. Originally Wegener had maintained that continents might float in the solid oceanic crust and that they might be dragged by tidal forces. Physicists scoffed that this was impossible, and of course this explanation was completely wrong. Nevertheless, Wegener was on the right track in terms of his geological observations.

From sea-floor spreading to plate tectonics

The breakthrough for understanding plate tectonics would not happen until the 1960s. Although not often realized, the development made use of improvements in our understanding of the interior structure of the Earth, which occurred in the first half of the twentieth century. Using seismic waves produced from natural earthquakes, scientists had already worked out that Earth had a metallic core, and that this was surrounded by a very thick mantle of dense rock. Small parts of the mantle were actually partially molten, and much of it was mechanically very weak; it could potentially deform or flow over long periods of time, despite being a solid. The Earth's crust, sitting on top of the mantle, was actually only a very thin shell – on average 40 km (25 miles) thick in continental areas and typically only 7 km (4 miles) thick under the oceans. This is a very small number given that the diameter of the Earth is nearly 13,000 km (8,000 miles)!

In compositional terms the Earth contains an iron-nickel core, a mantle made of relatively silica-poor rocks, known as peridotite, and the crust, which is broadly granite in the continents and basalt under the oceans. The outer layers of the Earth's structure can also be described in terms of their relative strength. The lithosphere is the rigid, brittle, outer shell of the Earth corresponding to the crust and upper few

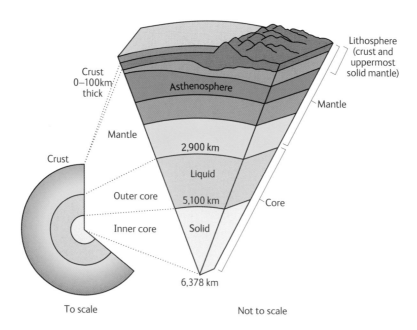

The structure of the Earth showing the Earth's relatively thin crust sitting on the mantle, which can deform or flow despite being solid.

kilometres of the mantle. The asthenosphere is the warm, mechanically weak part of the mantle which can flow and convect in a solid-state. The Earth's plates are made up of the lithosphere, riding on the weak asthenosphere.

Knowing that the rigid crust sat on a mantle which could potentially flow, as early as 1928 the British geologist Arthur Holmes suggested that perhaps convection forces in the Earth's mantle might drive the movement of the continents. The heat for this would come from the natural decay of radioactive elements in mantle rocks. But in the 1930s and 1940s we did not have satellite data to evaluate whether the continents were actually moving, so evidence would need to be sought elsewhere. A key discovery was sea-floor spreading. Bathymetric (i.e. sea floor) studies undertaken during the Second World War indicated that the mid-Atlantic actually had a ridge running along it, of elevated topography, where earthquakes associated with extension (rifting) could be found.

The oceanic crust here (and elsewhere) turned out to be made of black basalt, and not granite, which was associated with the interior of many continents. Further nautical studies showed that these mid-ocean ridges could also be located in the Pacific and Indian Oceans. The idea of extension (rifting) in these ridges was confirmed without a doubt in two main ways by the 1960s. First, it became clear that they were places where young basalts (i.e. young oceanic crust) were being continually produced from hot upwelling magma (molten rock, see chapter 2, Volcanoes). All the earthquake data suggested that the crust was being pulled apart in such locations. However, it was the study of magnetic anomalies in these young oceanic basalts at the seabed that sealed the deal. When igneous rocks cool below a certain temperature, magnetic minerals in the rock acquire a remnant magnetization that records the Earth's magnetic field at the time. We also know that the Earth's magnetic field reverses every few hundred thousand years, so igneous rocks show either positive or negative anomalies when measured. Good magnetometers (the instruments used for measuring this kind of signal) had already been developed in the Second World War – originally for detecting submarines, whose hulls were made of steel. So when geoscientists measured the magnetic field across the Mid-Atlantic Ridge, they made a critical discovery. There were stripes of positive or negative magnetization in the oceanic basalt, which ran parallel to the mid-ocean ridge.

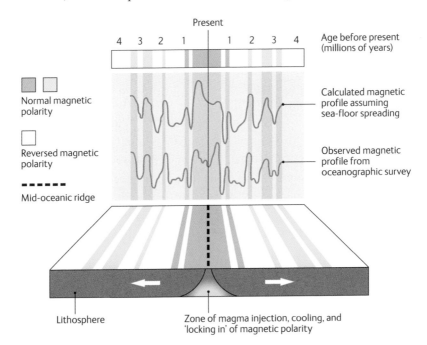

The Earth's magnetic field reverses every few hundred thousand years and is recorded in rocks. Identical reversals are found in basalts on either side of mid-ocean ridges showing sea-floor spreading.

What's more, the stripes of positive or negative magnetization were perfectly symmetrical on each side of the ridge and varied predictably, capturing the regular reversals of the Earth's magnetic field. This meant only one thing – new basalt (oceanic crust) was being formed at the mid-ocean ridge, and then it was pushed to each side as new rock was intruded. Over time new basalt was formed, sometimes with a reversed magnetization. As the process continued sea-floor spreading occurred. Geologists Frederick Vine and Drummond Matthews published these findings in a paper in the leading science journal, *Nature*, in 1963. Although their findings may seem technical in nature, it is worth thinking about the ground-breaking implications of this discovery. It meant that the Atlantic was getting wider by several centimetres a year! And it meant that the ocean crust near the mid-ocean ridge was young and warm, and the ocean crust nearer the American and Afro-European margins of the Atlantic was old, cold and dense (we now know this to be completely true). This old oceanic crust would have formed at the mid-ocean ridge tens of millions of years ago, when the Atlantic was far narrower! Here at last was the present-day observational evidence for Wegener's continental drift that matched his geological reconstructions.

There was one more piece of thinking that was needed to turn the observation of sea-floor spreading and continental drift into a coherent theory called plate tectonics. The argument went like this: if the oceans are spreading, does this mean that the Earth is actually getting bigger? How do you get long-term extension down the Atlantic, for instance, without increasing the circumference of the planet? In fact, in the 1960s, some geoscientists who worked on sea-floor spreading could not see a way round this problem and accepted that we must live on a growing Earth. But in fact the Earth is not getting bigger in size. Consequently, we need some way for oceanic crust to be recycled into the Earth's mantle, at a rate that balances the production of new crust. Was this something to do with the deep oceanic trenches found on the Earth? They were certainly areas where a lot of earthquakes and volcanoes occurred, and many of these seemed to be areas of contractional deformation. The final piece of the puzzle was the recognition that these deep trenches were areas where the process of subduction occurred. Here, old, cold and dense oceanic lithosphere (the crust and rigid upper few kilometres of the mantle) descend back into the deeper mantle, where they are eventually re-assimilated.

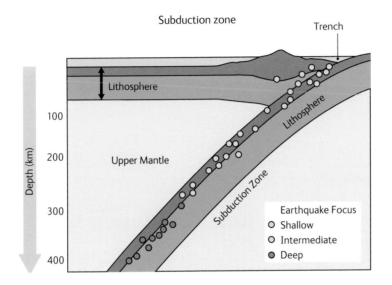

Subduction zone where cold brittle lithosphere, making the Earth's tectonic plates, is pushed into the mantle. Earthquakes can occur where the plate subducts.

Plate tectonics and plate boundaries

In its simplest form, the theory of plate tectonics holds that the Earth's lithosphere (i.e. the crust of the Earth and the mechanically strong upper few kilometres of the mantle) is not a continuous 'shell'. Instead, it is broken up into a number of rigid plates. These plates 'float' or slide over the asthenosphere – the part of the mantle that is warm enough that the rock it is made of has no mechanical strength; although it is a solid, over long timescales it can flow and deform. This is a bit like the way apparently solid resins, such as pitch, can flow if left for a long time. Scientists now believe that flow or currents in the mantle drive the motion of the plates. The largest plate by area is the Pacific plate; other plates include the North American plate, the Eurasian plate and the African plate. There are a number of smaller ones of varying size. It is important to realize that the plates do not stop where the continents stop. For instance, the North American plate is made up of the North American continent, and the oceanic crust right up to the middle of the Atlantic Ocean (the Mid-Atlantic Ridge). Fundamentally, plate tectonics works because the Earth's outer crust is strong, like the shell of an egg, but the asthenospheric mantle is weak and can flow. Plate tectonic theory suggests that the plates do not deform in their interiors. Instead they deform at their edges. To see plate tectonics in action we therefore have to visit plate boundaries. That is where we find phenomena like mountain belts, volcanoes, earthquakes and deep-sea trenches.

We can define three main types of plate boundaries, which reflect the relative movement of one plate relative to another:

1. Divergent plate boundaries, where tectonic plates move apart from each other and the plate area increases over time.

2. Convergent plate boundaries, where plates come together and the plate area is reduced. How the plates behave depends on whether the collision involves continental or oceanic crust, as we will see below.

3. Transform plate boundaries, where plates slide past each other and the plate area more or less remains the same.

The place where these boundaries meet is called a triple junction. Of course, there are some complexities involving the oblique motion of one plate relative to another, but these boundaries summarize the key types. Continental drift, as described by Wegener, is the result of the motion of the plates and these boundaries.

DIVERGENT PLATE BOUNDARIES

Divergent plate boundaries are typically characterized by oceanic spreading centres. It is here that two plates move away from each other, at a mid-ocean ridge. Hot molten rock, called magma, upwells, forming new oceanic crust. These spreading centres form a network that encircles much of the globe; we learnt about the magnetic anomalies recorded in the rocks near mid-ocean ridges in the previous section. These ridges have produced countless millions of square kilometres of oceanic crust in the past tens of millions of years. An important question to ask is whether divergent plate boundaries have to be under the sea. It generally seems that way when you look at the map of the plates. However, the story is a bit more complicated. Divergent plate boundaries can form in continental settings. These are known as continental rifts. If extension starts in the continent, and the two parts start to diverge strongly, a network of extensional geological faults can be formed. Further motion on the faults, typically causing

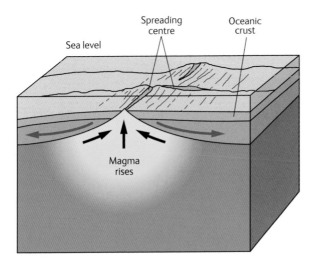

A divergent plate boundary in a mid-ocean ridge setting – magma rises producing molten basalt at the mid-ocean ridge.

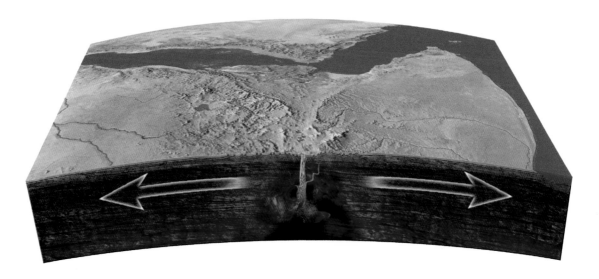

Continental extension (rifting) is occurring in East Africa. If this process continues a new ocean will be produced – this has happened in the Red Sea.

earthquakes, and the crust extends. And rather like pulling a piece of plasticine slowly apart, if you extend the crust you also thin it. Due to the reduced pressure, the warm mantle can upwell, and it can start to melt. Extensive volcanism can occur due to this. Sometimes the stretching and melting is so extensive that basalt is produced, much as we would find in the oceanic crust. If this process continues, a new ocean will be created as the old continental crust is thinned away, the centre of the rift falls below sea level, and young basalt is erupted at an incipient ocean spreading centre. In fact, this very event is currently underway and can be observed in the Red Sea and Afar region of Africa.

CONVERGENT PLATE BOUNDARIES

Convergent plate boundaries are where a lot of exciting geological phenomena happen on Earth. What is observed depends strongly on what types of plate are colliding.

1. If one plate is made of oceanic crust, and the other is a continental landmass, the plate comprising the oceanic crust is subducted; a deep linear oceanic trench may be found on the sea floor where this is occurring. This happens because the continental crust is made of granite, which is lighter and more buoyant than the oceanic plate, which is made of more dense basalt. The down-going plate (sometimes called the 'slab') is recycled into the mantle, but it may descend for hundreds of kilometres, and may take millions of years to be re-incorporated with the rest of the mantle. Water that is present in the subducted plate, for instance in marine sediments lying on top of the original oceanic crust, is driven off as the slab gets deeper and is heated up. The addition of this fluid to the rock above the down-going plate can promote further melting, which can lead to the development of explosive volcanism in the over-riding continental plate, with magmas rich in water (steam) and gases being erupted. Sediments lying on top of the down-going plate can also be scraped off and squashed together, forming a wedge of highly deformed material that may partially fill the oceanic trench. A belt of mountains forms parallel to the subduction zone as the continental plate is deformed, thickened and shortened. Large earthquakes occur frequently because of the build-up of frictional stress at the plate boundary as the slab is pulled down into the mantle. A classic example of this kind of plate boundary is the western margin of South America, where the Nazca plate subducts beneath the South American plate.

2. If both plates are made of oceanic crust, one of them eventually subducts; usually this is the one that contains oceanic crust that is older, colder and denser. Again, as subduction occurs, water can be driven off the descending slab, creating melting and volcanism. There is no continental landmass present, so instead of a mountain belt, a line of volcanic islands – known as an island arc – is formed. A classic example is where the Pacific plate is subducting to the west underneath the Philippine plate. This is where the famous Marianas trench, at a depth of 11 km (7 miles) is located!

3. The third and final combination is where we find two plates with continental landmasses colliding. Continental crust is very difficult to subduct on account of its relatively low density compared to mantle rocks,

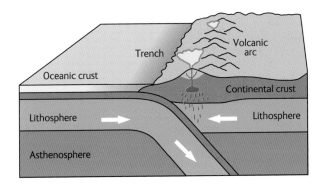

A convergent plate boundary can occur where one plate subducts under a continental landmass (**LEFT**), such as the Pacific crust subducting under South America. An oceanic trench mountain range (the Andes) and volcanoes are generated. Sometimes two continents collide, buliding a large mountain range and a high plateau, such as the Himalayas (**BELOW**).

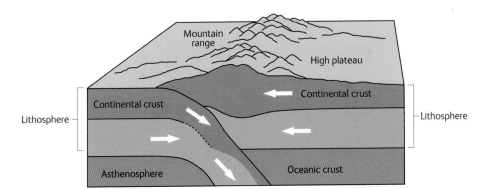

so one plate ends up being thrust above the other. A large, extensive and high elevation mountain range forms, often with a wide plateau. The crust becomes highly thickened in this process. Complex geological structures, including folds and faults form, and parts of the crust and sedimentary units caught up in the collision can be buried to great depths and undergo extensive metamorphism. Melting of the crust at depth may lead to the formation of igneous rocks such as granite, which are common in this type of setting. Again, earthquakes are also very frequent in these regions. The most famous example of this kind of continental collision is the Himalayas and the Tibetan plateau. You may think of Mt Everest as being a permanent feature on Earth. But it did not even exist 50 million years ago. In fact, in the Late Cretaceous, when dinosaurs still roamed the planet, India was approximately 6,500 km (4,000 miles) south of Asia, somewhere in the middle of what is now the Indian Ocean. Many of the deformed rocks found in the Himalayas record the history of the convergence of the Indian plate with Eurasia. Marine limestones more than 400 million years old are found right at the summit of Everest at nearly 9 km (5½ miles) elevation, and they have been uplifted from below sea level during this great mountain building process.

TRANSFORM PLATE BOUNDARIES

Boundaries where plates slide horizontally past each other – without extension or subduction – are known as transform plate boundaries. They can occur on the continents, most famously the San Andreas Fault in California (which is the transform plate boundary between the North American plate and the Pacific plate). These continental plate boundaries can give rise to large earthquakes, as the plates jam up against each other, before suddenly moving. Complex geological structures can be created where the transform boundary is not aligned perfectly with the direction in which the plates are moving. We also see these kinds of transform plate boundaries prominently in the oceans. For instance, the Mid-Atlantic Ridge that we have already discussed has many transform faults that off-set the oceanic spreading centres. These accommodate lateral movement as the North American and African plates move apart, and can be clearly seen from bathymetric (seabed) surveys.

A transform plate boundary occurs where the plates slide past each other (RIGHT), such as is happening today in the San Andreas Fault. Transform faults also offset mid-ocean ridges under the sea (BELOW).

Plate tectonics present, past and future

Today we know from satellite observation that the plates are moving at rates of millimetres to centimetres a year. Not all plates move at the same rate relative to each other; for instance, the Atlantic is opening at around 5 cm (2 in) a year. At the mid-ocean ridge boundary in the Pacific (called the East Pacific Rise) the Pacific and Nazca plates are actually moving away from each other at approximately 15 cm (6 in) per year. Near the island nation of Tonga, the Pacific plate subducts under the Australian plate at nearly 25 cm (10 in) a year. This amounts to nearly 30 m (100 ft) of shortening since the beginning of twentieth century alone. Scientists do not fully understand what controls these differences in rate, but it may relate to the vigour of convection and flow in the Earth's mantle driving the plates. It may also be due to particular geological circumstances such as the amount of the plate boundary that is surrounded by subduction zones, since the subducted part of the plate may 'pull' the rest of the plate at a faster rate, driving faster spreading at the mid-ocean ridge. The rates of motion we are talking about may not seem that rapid, but in a geological context, where we comfortably talk of millions or tens of millions of years, and where the Earth is more than 4,500 million years old, the numbers are huge. At these rates, it is easy to see how whole oceans could be created and destroyed over geological time periods, and indeed, we now know that is indeed what has happened in the past. In fact, the motion of the Earth's plates, including the birth and creation of oceans and mountain belts, explains the complicated geology we can observe right around the world. As we have seen, it also explains where and why we see earthquakes and volcanoes today, and a wide range of other physical phenomena and observations in the Earth sciences. It is for this reason that plate tectonics is considered the unifying theory in geology, in the same way that Darwin's theory of evolution explains much about the field of biology, and Einstein's theory of relativity explains much about physics.

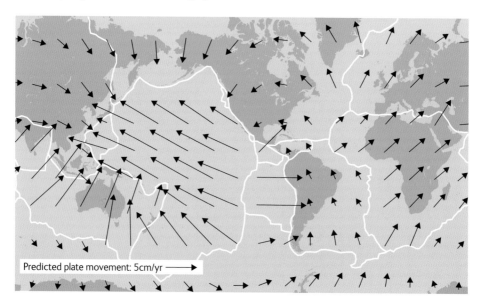

Predicted plate movement: 5cm/yr ⟶

Movements of the tectonic plates taken from GPS data. The length of the arrow is proportional to the speed the plate is moving.

So can we run plate tectonics backwards, and see where the continents and oceans were in the past? Excitingly, the answer to this question is yes. The first place to start is in the oceans. Thanks to magnetic anomalies and other measurements, we know the age of oceanic crust quite well. We can actually produce maps of this across the globe (see below). Red colours symbolizing young crust are found on either side of the mid-ocean ridge; colder colours (representing older crust) are found further away. We can therefore attempt to restore the past position of continents based on this information (along with some other geological constraints, too, of course). One catch for going further back in time is that there is not much oceanic crust older than the beginning of the Cretaceous period at 145 million years ago. This is because the older ocean floor has already been completely subducted. So for time periods older than this, we have to rely on geological, geophysical and palaeontological evidence. For instance, using the types of evidence that Wegener originally put together to argue for continental drift in the first place, we can try to re-assemble

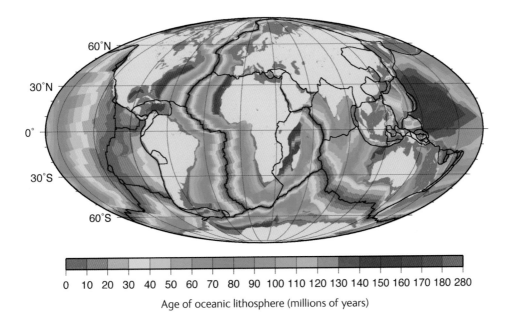

Age of oceanic lithosphere (millions of years)

Global map of the age of the ocean floor based on the patterns of magnetic field intensity reflecting the age of the oceanic crust.

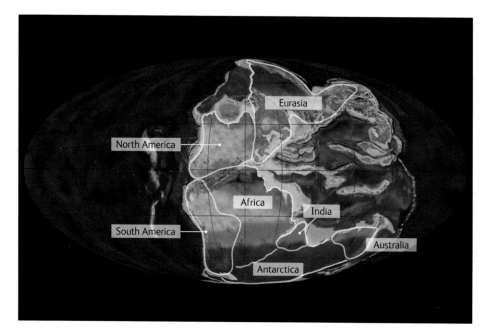

Plate tectonic reconstruction of the location of Pangea, which existed about 250 million years ago, showing how today's continents formed the supercontinent.

former continental landmasses on the basis of similar geology, structure and fossil evidence. The most famous of these is the supercontinent of Pangea, which existed about 250 million years ago at the end of Permian period. This supercontinent was surrounded by a large ocean called Panthalassa. The Pacific today is a smaller portion of this once even larger ocean. Pangea broke up when the Atlantic opened, rifting North and South America away from Africa and Eurasia.

Further back in time still, we have evidence for other ancient supercontinents, and other times when the continents were separated. We can use ancient mountain belts, such as the Highlands of Scotland and the Appalachians in eastern North America, to reconstruct where continental collisions took place in the past, and we can use remnant magnetization in rocks (in a similar way to the magnetization of basalts on the ocean floor) to determine at which latitude the rock was formed. In this way, we have good evidence to suggest that England and Scotland were attached to separate continental landmasses that collided about 400 million years ago. This collision happened while Britain was in the southern hemisphere, at about 40 degrees south. Britain is currently 50 degrees north meaning that its continental landmass has drifted by thousands of kilometres over hundreds of millions of years. It is likely that this repeating plate tectonic cycle of continental rifting, drifting and collision – known as the Wilson cycle – has taken place a number of times in Earth's history.

One important thing to recognize is the crucial role that continental drift has had on the environmental, biological and physical world around us. The motion of Antarctica towards the South Pole since the Cretaceous, and the development of the circum-antarctic polar current promoted the development of the polar ice caps and ice sheets that play a crucial role in the Earth's climate today. The rifting of islands such as Australia and Madagascar away from other landmasses promoted the evolution of unique indigenous flora and fauna. The rise of the Himalayas generated the monsoon that dominates the climate and hydrological cycle in India and Pakistan by creating a steep-fronted, high mountain belt that deflects winds from the south upwards, and promotes

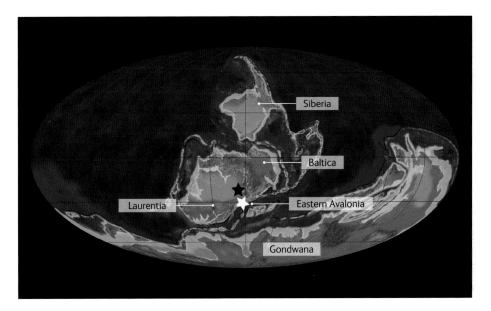

Plate tectonic configuration for the end of the Ordovician, approximately 445 million years ago. England (white star) was part of small continent (Eastern Avalonia) that had collided with Laurentia and Baltica, where Scotland was located (black star).

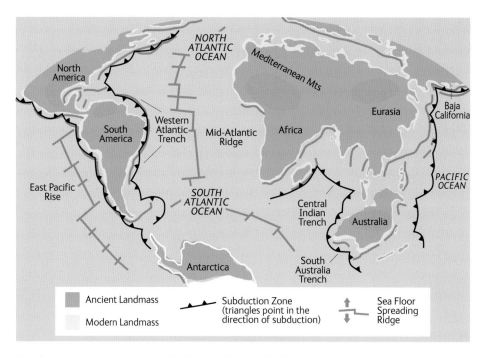

How the tectonic plates may look in 50 million years' time, with Africa moving northwards fully closing the Mediterranean Sea, and North and South America continuing to move away from Africa and Eurasia.

rainfall. And the migration of landmasses over thousands of kilometres across the equator explains how 250-million-year-old desert rocks can be found in the temperate, rain-soaked British Isles. These are just four examples of the way in which plate tectonics have influenced our planet's history, as well as its present.

What about the future? Predicting continental configurations in millions of years' time is tricky because we cannot foresee every new geological development. But we can make predictions, and some things look to be very clear: North and South America will likely continue to move away from Africa and Eurasia; Africa is likely to continue to move north, fully closing the Mediterranean Sea; Australia will move to the north, eventually colliding with Southeast Asia; and California will be moved northwards relative to the rest of America, ending up somewhere near Alaska. A future world in 50 million years will look very different to now.

Natural hazards and plate tectonics

We live on a restless planet and an increasingly crowded world. The natural hazards of Earth, such as earthquakes, landslides and volcanic eruptions, challenge societies and governments worldwide. What is less often appreciated is that plate tectonics plays a crucial role in determining the frequency, magnitude and location of these hazards. A quick glance at the map tells us that billions of people across the globe live near a plate boundary and, as we have seen, this is where much of the geological action happens. Volcanic eruptions, as explained in the following chapters, are one famous hazard associated with subduction zones, where explosive volcanism arises as a result of hydrous melting above the subducting slab. The distribution of numerous volcanoes around the Pacific Ocean – known as the Pacific Ring of Fire – highlights the close relationship between plate tectonic boundaries and volcanism. The ring is more than 40,000 km (25,000 miles) long and is associated with a nearly continuous series of subduction zones along the margins of the Pacific. It contains more than 450

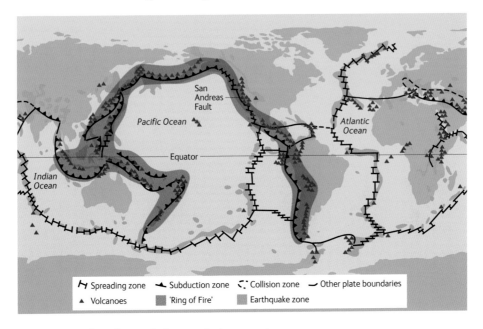

The Pacific Ring of Fire showing the location of volcanoes and earthquakes around the Pacific. All magnitude 9 earthquakes since 1950 have happened on subduction zones forming this Ring of Fire.

volcanoes, which comprise 75% of the active or potentially active volcanoes in the world today. In fact the majority of major eruptions occur on volcanoes linked to this significant geological feature.

Of course, at convergent margins where subduction occurs, earthquakes are also a major problem (see chapter 3, Earthquakes). Indeed, the largest earthquakes the world has known, registering more than magnitude 9 on the moment magnitude scale have occurred on these plate boundaries. Big earthquakes occur here because the rate of convergence between the subducting plate and the overriding plate is usually fast, meaning a rapid build-up of stress. However, the plate boundary can be thousands of kilometres long, and the plate surface can be very rough; these factors promote the frictional 'sticking' of the plate interface. When the stress build-up reaches a tipping point, the subducted plate finally moves. The huge energy released generates seismic waves that travel through the crust. A magnitude 9 earthquake releases energy equivalent

OPPOSITE AND ABOVE: Earthquake damage from the 2011 Tōhoku earthquake in Japan, a result of the Pacific plate subducting westwards under the Japanese archipelago.

Damage from the 2004 Indian Ocean earthquake resulting from the subduction of the Indian plate under Indonesia.

to approximately 2 trillion kilograms of explosive, such as TNT. Tsunamis, fast-moving waves created by the motion of the seabed displacing the water column, can be generated. All five of the magnitude 9+ quakes documented since 1960 have occurred on convergent plate boundaries. Well known examples include the 2011 Tōhoku earthquake in Japan, at magnitude 9.1, where the Pacific plate is subducting westwards under the Japanese archipelago; and the magnitude 9.3 Indian Ocean earthquake in 2004, where the Indian plate is subducting under Indonesia. The Japanese quake and associated tsunami killed nearly 20,000 people, and promoted the infamous nuclear meltdown at the Fukushima power station. However, the Indian Ocean quake killed a quarter of a million people in over 14 countries. It was one of the deadliest events of the last 500 years, primarily because of the large tsunami that travelled across the globe. The location and severity of these incidents is pre-determined by the planet's plate tectonic configuration. In the future, geologists believe that 'megaquakes' will occur on other large subduction zones, such as where the Nazca plate subducts underneath South America. Elsewhere, in areas of active continental rifting and extension, earthquakes are generally smaller as the size of the faults in these cases are not as big. However, they can still be devastating, especially where they occur at relatively shallow depths. It is also important to realize these earthquakes can promote other hazards – for instance the Tōhoku earthquake promoted thousands of landslides and debris flows in the steep mountainous topography associated with the collision of the Pacific and Eurasian plates. These events claimed additional lives and blocked access to towns and villages. Moreover, at a later date storms can mobilize the sediment generated by the earthquakes, which can raise riverbeds and promote flooding. Understanding and improving our resilience to this 'hazard cascade' is a major challenge for Earth scientists today, and our ability to address these questions must be rooted in understanding the plate tectonic context in which they occur.

2 Volcanoes

On 20 February 1943, Dionisio Pulido and his family were working in their cornfield near the city of Uruapan, in the Mexican state of Michoacán, about 320 km (200 miles) west of Mexico City. For weeks there had been tremors and deep rumblings from within the Earth. At 4pm a crack opened in Dionisio's cornfield and he watched in amazement as the ground rose up about 2 m (6½ ft), emitting ash and gas. There were loud explosions and a strong smell of rotten eggs (a clear sign of hydrogen sulfide emission). A new volcano had been born! Within hours the fissure was transformed into the small crater of the newly created Parícutin volcano. After 24 hours, the cone was 50 m (164 ft) high, and in a week it reached about 150 m (492 ft). The activity of Paricutin lasted for 9 years, forming a 424 m (1,391 ft) high cone and completely covering the nearby village of San Juan Pararingutíro in lava and scoria – a dark rock, relatively light due to the presence of lots of holes, a bit like a pumice, which is formed during volcanic eruptions that emit low silica magma. In the end the eruption affected an area of 233 km² (90 sq miles) and forced hundreds of people to be relocated from their homes.

OPPOSITE: Lava fountain from one of the craters of Mt Etna, east coast of Sicily, Italy.

LEFT: The village of San Juan Pararingutíro, Mexico was completely covered in lava from the activity of Parícutin volcano – seen in the background – over a period of 9 years.

Volcanoes have fascinated humans since the beginning of our journey on Earth with their terrifying beauty and their power of destruction. The earliest humans must have known, too, that the lands around volcanoes were rich and fertile. Volcanoes play such an important role in life on Earth, even for those who have never stepped foot on one. It is no coincidence that our early ancestors evolved in Africa in the volcanic region of the East African Rift. Active volcanoes, of course, pose a threat for those living nearby, but life on Earth would have been impossible without the key role played by volcanoes in the development of the atmosphere and fertile soils, among many other benefits.

In this chapter, how volcanoes form and the different types of activity is discussed. The benefits and dangers that they present to humans is discussed in greater detail in Chapters 4 and 5, together with the impact of volcanoes on climate and the environment.

What makes a volcano?

Volcanoes are not randomly distributed. They are concentrated in certain areas of the Earth, as becomes clear when you look at their distribution on a map. This distribution tells us that they are linked to plate tectonics (see chapter 1,

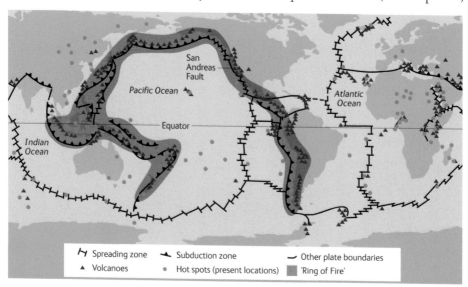

Distribution of volcanoes around the world showing the Ring of Fire encircling the Pacific Ocean.

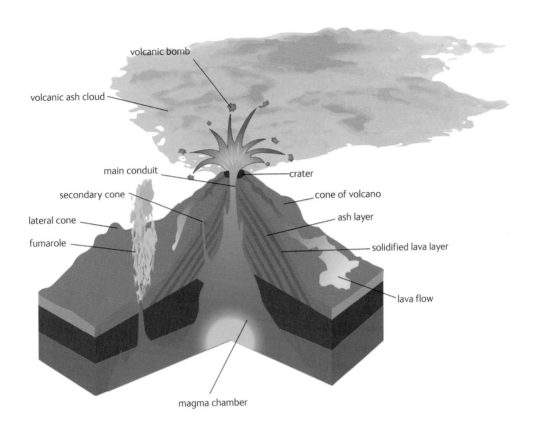

volcanic bomb

volcanic ash cloud

main conduit

secondary cone

lateral cone

fumarole

crater

cone of volcano

ash layer

solidified lava layer

lava flow

magma chamber

The schematic internal structure of a volcano showing the key features relevant for many types of volcano.

Tectonics) and we find them where two plates collide, as seen in the so-called Ring of Fire that encircles the Pacific Ocean. They also occur where two plates separate, forming the mid-ocean ridge or a continental rift, if the two separating plates are continental. Finally, volcanoes can also be located in the middle of plates associated with mantle plumes, which are concentrated regions where magma rises from the deep mantle to shallow depths. Such regions are also called hot spots, a classic example of which are the Hawaiian volcanoes.

A volcano is usually high ground or a mountain, sometimes with very steep flanks, produced as a result of the eruption of magma, a mixture of molten rocks and gases, outpoured in a gentle way or in an explosive fashion from an opening in the Earth's crust. A volcano is a complex system that can be described in three parts: the volcano edifice, usually with a crater area on top, where lava, ash and gases are emitted; a conduit, where magma rises to the surface; and a plumbing system, where the magma is stored.

What makes the magma?

Magma is a very hot mixture of molten rock, gas, and crystals, sometimes carrying also fragments of solid rock. The temperature is usually around 800–1,200°C (1,472–2,192°F). Once erupted, magma is called lava and upon cooling, it solidifies to form glassy or crystalline igneous rocks. Magma is prevalently composed of silica, the oxide of silicon (SiO_2) formed by the two most abundant elements in the Earth's mantle and crust, silicon (Si) and oxygen (O). Other important elements such as aluminium (Al), magnesium (Mg), iron (Fe), calcium (Ca), sodium (Na), potassium (K), titanium (Ti), and many more, are present in various proportions. Given that silicon is the most abundant element in the Earth's mantle and crust, most magma and the rocks that are formed upon magma solidification are called silicates. Two of the most common rocks formed from the solidification of magmas (so called igneous rocks) in the Earth's crust are basalt, a volcanic rock with a relatively low silica content, and granite, a plutonic rock (cooled beneath the Earth's surface and rich in crystals) with a higher silica content than basalt. There are some unusual non-silicate magma compositions such as carbon-rich magma (carbonatite) that erupts in certain volcanoes, such as the Ol Doinyo Lengai volcano in Tanzania, and sulfur-rich melts outpoured by some fumaroles. However, these are both rather rare and the prevalent composition of magmas is silicatic.

Under the Earth's brittle surface, there is no such a thing as a magma layer or 'ocean' from where the magma is siphoned off to the surface. Instead, partial melting of crystalline rocks, at depth, generates magma. But how can a rock melt? Temperature and pressure increase with depth from the surface of the Earth towards the core. The rate at which the temperature increases with depth is about 20–25°C/km, for the first 100 km and this increase is known as the geothermal gradient. There are some exceptions to this, such as hot spots and areas of thin crust where the geothermal gradient is higher, 30–50°C/km, as well as areas where the geothermal gradient is lower, 5–10°C/km as in subduction

Two of the most common igneous rocks (formed from the solidification of magma) are basalt (TOP), this one from Hawaii, and granite (ABOVE). The granite, a crystalline rock formed by magma solidification under the Earth's surface, contains white grains of feldspar, clear grains of quartz and black grains of biotite or amphibole.

zones and areas with thickened crust. Despite the increase of temperature with depth, rocks stay solid because the higher the pressure the greater the melting point of the rock. As shown in the graph opposite, the geotherm (the rate at which temperature increases with depth) and the rock's solidus line (the temperature above which the rock melts) do not cross each other under normal conditions. However, at depth the Earth's mantle is not rigid like the overlying lithosphere (the outer shell of the Earth formed by the crust and the upper mantle with a variable thickness up to about 280 km), but ductile. It can deform plastically and is called the asthenosphere (i.e. ductile asthenosphere) (see p.14). When two crustal plates are pulled apart as in the oceanic basins, the released burden allows the asthenosphere to rise upwards filling the gap in the rift. This decompression favours spontaneous melting of the mantle, because the decrease in pressure is quicker than the cooling of the rock. Thus, the rock is hotter than it would be in relation to the lower pressure, and it melts (the vertical arrow in the graph). Decompression melting is the most common melting process in the Earth and produces all the oceanic floor basalts. It is also the common mechanism that produces magmatism linked to a hot spot, such as Hawaii, where the hot plume of mantle rises to produce a chain of volcanic islands.

There is a second common mechanism that allows the mantle to melt and thus cause the production of magma. This is shown in the graph by the wet solidus curve. The melting temperature of rocks is decreased by the addition of water, thus the solidus line moves towards the geotherm. When they cross, the rocks start to melt. Depending on the amount of added water, the exact position of the crossing point changes. This second melting mechanism is commonly seen in a subduction zone, where water is brought to depth by the subducting slab. Slabs are composed of oceanic sediments and basalts. Both contain minerals that can contain some water and other fluids, as carbon dioxide (derived from carbonates), in their structure. When heated, these minerals can release the water, along with other elements, while the slab is pulled down into the mantle.

Finally, the third melting mechanism is simply to heat up the rocks to their melting point. The heat is usually provided by hot basaltic magma ponding at the base of the crust due to its higher density. Large volumes of silica-rich rocks (rhyolites) can be generated in this way by the melting of the thick continental crust in some cases. Basaltic rocks are by far the most common rocks and basaltic magma produced by melting of upwelling parts of the mantle is by far the most common type of magma on Earth.

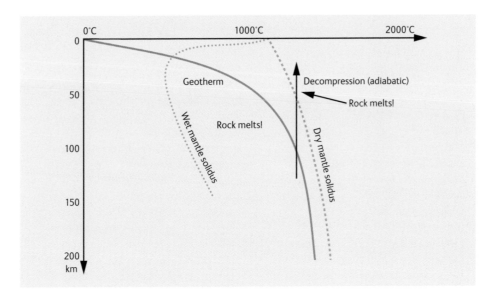

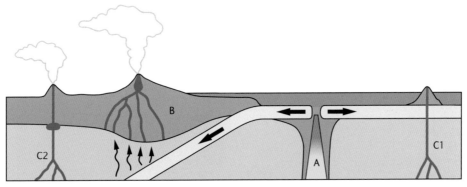

TOP: The Earth's temperature increases with depth, with a rate shown by the red line (geotherm) – this rate is called the geothermal gradient. The solidus (the temperature above which the rock melts) and the geotherm do not cross each other under normal conditions and the temperature at depth is not enough to melt dry mantle rocks (brown dotted line) and the mantle is solid. Mantle rocks can melt via decompression (black vertical arrow) or if water is added via subduction (blue dashed line showing the wet mantle solidus).

ABOVE: Magma generation mechanisms. A, decompression melting at a mid-ocean ridge; B, melt production at a subduction zone favoured by the addition of water (vertical arrows) from the subducting slab; C, decompression melting associated with mantle plumes at a hot spot (C1) and intra-plate setting (C2).

A VOLCANO'S PLUMBING SYSTEM

We imagine a volcano to have a pond of magma beneath the surface, the so-called magma chamber. Images on the internet show volcanoes with a balloon-shaped magma chamber filled by liquid magma feeding the lava flows, and the explosive activity taking place at the top of the volcano, (i.e. the crater). The concept of a magma chamber as a large pond of fluid magma is convenient and has been used for more than 100 years. It explains many processes taking place, such as different eruptive styles, the large variability in the chemical compositions of volcanic rocks, volcanic earthquakes and gas emissions. However, the increasing sophistication of scientific instruments and analytical techniques has led to a better understanding of the volcano's plumbing system. Combining observations from different fields – geophysics, volcanology, geology, geochemistry, petrology – has led to the idea that volcanoes are fed by a storage region that is very complex and extends vertically into the lower crust for several kilometres. It is not a continuous, but rather a complex discontinuous system where the magma is accumulating at various levels. The concept of a melt-dominated magma chamber is now radically changing to that of a solid-dominated 'mush' system, where the mush indicates a continuous framework of solid crystals where the melt is only a small fraction distributed through the solid portion.

ABOVE: Melt-dominated magma chamber.
RIGHT: Complex mush-dominated vertically extended plumbing system – a transcrustal system.

Lava flows and big explosions

When we imagine a volcano, we usually think of a big explosion with incandescent ash, gases and rocks spewing into the air and a big umbrella-like plume. Certainly, this is the most spectacular type of eruption, but there are many other types of eruptions. These include the lava flows and spectacular fire fountains of Etna, Hawaii and the recent Bárðarbunga Icelandic lava flow. Volcanic eruptions can be divided into two main groups: effusive and explosive.

Lava fountaining at Stromboli volcano, off the north coast of Sicily, Italy.

Lava flow from a fissure eruption at Bárðarbunga, Iceland.

EFFUSIVE ERUPTIONS

Lava flows and sticky lava that forms bulging domes belong to the effusive type of eruption. There is very little gas, therefore very little or no explosion is involved in the outpouring of the magma. Hawaiian and Icelandic volcanoes are mostly characterized by effusive eruptions, but whenever lava flows this can be considered an effusive eruption. Many explosive volcanoes also have lava flows, along with explosive eruptions. Basaltic lavas are fluid and they cool down to a shiny smooth surface called *pahoehoe* (a Hawaiian name which means 'on which one can walk'). Upon cooling the surface can form wrinkles and it is then known as ropy lava. The surface of pahoehoe lavas cool very quickly and can form an external carapace under which the lava flow continues to travel over tens or hundreds of kilometres in lava tubes. These form when the external

Basaltic lavas are fluid and they cool down to a shiny, smooth surface called pahoehoe, like this lava flow on the island of Hawaii.

ABOVE: On cooling basaltic lavas wrinkle and are known as ropy lavas like these on Rabida Island, Galapagos.

RIGHT: Lava can continue to flow under an external cooled surface in lava tubes such as this tube at Stromboli volcano, Italy.

surface cools down but the interior is still hot and keeps flowing, thus the external cooled surface forms a sort of tube. When the eruption finishes, the drained lava tube can survive as a spectacular tunnel. Another common type of lava flow is called *aa*, another Polynesian name. It is very common on both oceanic islands and continents and its name is due to the very rough, blocky and sharp surface, which makes it very difficult to walk on. Most pahoehoe lava, upon cooling but before solidifying, becomes *aa* lava because of the increased viscosity (or stickiness) due to the cooling process.

Spiky, uneven aa volcanic lava flows, from the Old Peak (Pico Viejo) of Mount Teide, Tenerife, Canary Islands.

Thick lava or huge flood basalts can cool down and fracture on the surface forming columnar hexagonal shaped basalts like the Giants Causeway, Northern Ireland.

When lava flows interact with water, as for example from an effusive eruption under the sea at the mid-ocean ridge or when the lava flows into the sea, pillow lava will form. A pillow is a round to elongate kidney-shaped basaltic body. The water cools the exterior of the lava very quickly in a sort of bulge or pillow-shape, while the interior is still hot. The continuous flowing of the lava breaks the pillow at one point allowing a second pillow to form. This, in turn, will be broken by the flowing lava and a third pillow will form attached to the second. The process continues until the end of the eruption. The result is a lava flow formed of densely packed individual pillows, each one attached to the other.

When thick lava flows or huge flood basalts cool down, fractures form on the surface and propagate inwards into the thick slowly cooling lava, forming columnar hexagonal shaped basalt. The famous Giant's Causeway in Northern Ireland is an impressive example of columnar basalts formed in this way by a volcanic eruption that occurred around 50 to 60 million years ago.

EXPLOSIVE ERUPTIONS

Around 67 large cities (>100,000 inhabitants) are located on or close to an active volcano – one that has the potential to erupt either effusively or explosively or both. Among those, there are three megacities: Tokyo in the shadow of Mt Fuji; Manila close to Pinatubo volcano and Mexico City, which is not far from the towering Popocatepétl volcano. All three volcanoes are strato- or composite volcanoes that have the classical steep cone shape. They are also very explosive volcanoes. The majority of explosive volcanoes are located along the Ring of Fire, above subduction zones. However, other types of volcanoes can also generate explosive eruptions.

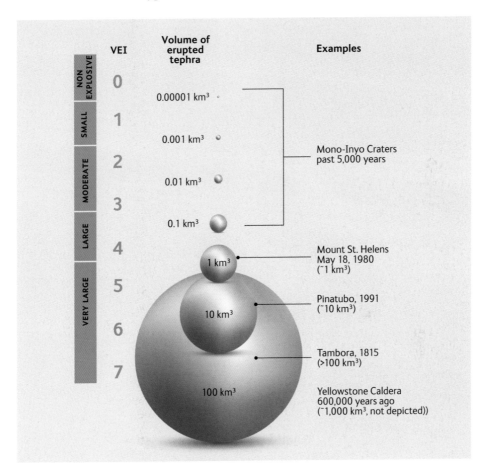

Volcanic Explosive Index (VEI) of some famous eruptions is shown alongside the volume of tephra emitted during the eruption.

When measuring the relative size of a volcanic eruption, two factors are important: the eruption column and the explosiveness. The height of the eruption column can be measured directly in observed eruptions. The extent of the dispersion of tephra (see pp.56–57) around the volcano, whose volume is correlated with the size of the eruption, can be used to estimate the size and height of the eruption column, when the eruption is not directly observed. The explosiveness of a volcano is expressed by the Volcanic Explosive Index (VEI). This is a sort of intensity scale, similar to the one used for earthquakes, which takes into account a number of criteria to estimate the force of the volcanic event. The VEI runs from 0 for not-explosive or very small explosive eruption (called Hawaiian) up to 8 for the colossal ones (termed Ultra-Plinian). The VEI can also be linked with the volume of emitted material, to give a better idea of how big the eruption is, and is also loosely linked with the frequency of the eruption. Larger eruptions are less frequent than smaller ones, which is very good news for all of us.

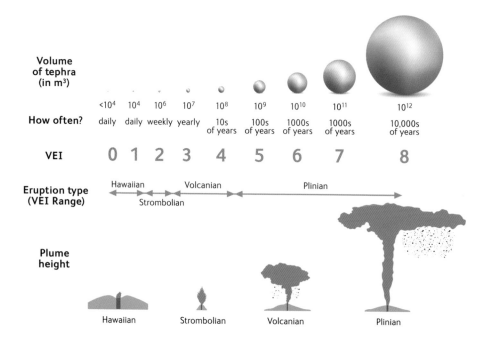

Volcanic Explosive Index (VEI) with the volume of erupted material, relative frequency of eruption and plume height for different types of volcanic eruptions.

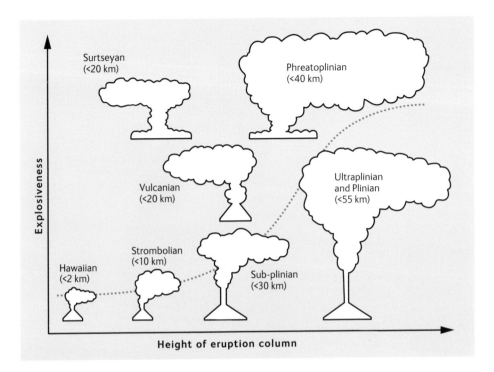

Classification of explosive eruptions.

At the low intensity end of the VEI scale we find Hawaiian eruptions, named after the style of activity currently observed in Hawaiian volcanoes. They are characterized by lava fountains that can be quite spectacular with hot, incandescent clots of magma ejected at high speed (up to 100 m/s) from a vent, typically rising to a height of a few tens to a few hundred metres before landing back on the ground. The emitted magma is still extremely hot (about 1,100°C or 1,832°F) and if the fountaining activity is very intense, when the erupted material falls on the ground it can accumulate to form lava flows, as frequently observed on Hawaiian volcanoes. Any type of activity similar to this one is called a Hawaiian eruption regardless of where in the world it occurs.

Stromboli volcano in southern Italy is considered the lighthouse of the Mediterranean because of its constant activity, with minor eruptions that are visible from many points in the surrounding sea. Indeed, Strombolian eruptions consist of transient explosions of short duration that occur in sequence at intervals of a few minutes to a few hours (frequently exploding every 20 minutes

A moderate explosive eruption at Kagoshima volcano, Japan.

or so). Each explosion generates a small plume of some hundred metres in height throwing ash, incandescent volcanic bombs and large ballistic blocks. Apart from Stromboli, there are many volcanoes that show the same Strombolian-type of activity with transient, short yet regular minor eruptions. Both Hawaiian and Strombolian eruptions are characteristic of basaltic magmas.

The famous 79 AD Pompeii eruption of Mt Vesuvius in southern Italy caught the local population by surprise. The vivid account by Pliny the Younger, a Roman administrator and nephew of Pliny the Elder a naturalist, philosopher and author of the 37-volume *Naturalis Historia*, who lost his life during this eruption, gives a clear picture of the devastating consequences of the unexpected large explosive eruption. We now have a much better knowledge of volcanic behaviour and can recognize the signals of unrest that precede an eruption, but at the time the long period of inactivity of Vesuvius gave a false sense of security. The eruption

TOP: The bay of Naples with Mount Vesuvius covered in snow and erupting 6 January 1836 (a gouache painting by Mauton, 1836).
ABOVE: Vesuvius and Mt Somma with the now large and overpopulated city of Naples at the bottom of its flanks.

started on the morning of the 24 August with a volcanic plume rising above the vent up to 33 km (20 miles) and a dense rain of ash and pumice, which was not necessarily lethal. Indeed, the devastation arrived around midnight when the eruptive column started to collapse generating the first devastating pyroclastic flow, an avalanche of hot ash, pumice, rock fragments and volcanic gases. This first pyroclastic flow rolled down the flank of Vesuvius at a speed of up to 100 km/h (62 miles/h), wiping out everything in its path including the city of Herculaneum on the coast. The city of Pompeii suffered a similar fate only a few hours after Herculaneum, due to a second pyroclastic flow. A layer over 3 m (10 ft) thick of hot volcanic material covered the two cities and the devastation can still be seen today after the archaeological excavation revealed the casts of thousands of human bodies. It is unknown exactly how many died but it was a catastrophe. Most of the people died due to suffocation by volcanic gases and heat and they are eerily preserved as they were in their last moments.

People engulfed in Pompeii by the pyroclastic flows from the eruption of Vesuvius, Italy in 79 AD.

Eruptions like the one at Pompeii are called Plinian eruptions, named after Pliny the Younger, and characterized by a jet of magma and volcanic gases emerging at high speed – about 100–600 m/s (224–1,342 miles/h) from the vent and forming the classical umbrella-shaped head eruptive column. The eruption can last for hours or days. The convective plume sucks up the surrounding air into the jet and expands into the atmosphere reaching a height of about 55 km (34 miles). Depending on the intensity of the eruption, Plinian eruptions are also subdivided into Ultraplinian if they are larger than Plinian and Subplinian if the eruption is slightly smaller.

When magma interacts with water, eruptions get even more explosive and are called hydromagmatic or phreatomagmatic. Magma can interact with water in a wide range of environments such as seawater, lakes and glaciers. The island of Surtsey off the south coast of Iceland, was born between 1963 and 1965 due to the interaction of a submarine volcano and the shallow marine environment. The eruption started on 14 November 1963, when the top of the volcano was about 10 m (32 ft) below the water surface. A dense black cloud of ash and steam rising in a few hours to 65 m (213 ft) above the sea surface was the first sign of the eruption. By the next day, Surtsey was well above sea level and continued to grow. This type of eruption has been known as Surtseyan ever since.

What makes a volcano explode?

Volcanoes erupt explosively because of the gases (mainly gaseous water and carbon dioxide) dissolved in the magma. However, the amount of dissolved gases (also called volatiles) depends on a variety of factors among which pressure and magma composition, particularly viscosity, are the most important. When the magma rises towards the surface, pressure decreases and the gases tend to escape. Bubbles begin to form in a process called vesiculation or gas exolution. The bubbles start growing and coalesce together leaving the magma in the interstices between the bubbles. This process is called fragmentation because the magma is fragmented by the growing bubbles. At this point the explosive eruption will occur. In this way, the amount of dissolved gases represents the driving force of an explosive eruption. However, the viscosity of the magma is equally important in determining if the eruption will be explosive or not. Indeed, in low viscosity magma, like basalt, the gas can readily escape producing an effusive or low explosive Hawaiian eruption. At the other extreme, high

ROCKS FORMED DURING EXPLOSIVE ERUPTIONS

Rocks generated during an explosive eruption are called pyroclastic rocks or tephra. These are general terms to indicate any loose deposit made of volcanic material, independent of size and composition. If a fine-grained deposit is consolidated and hardened (i.e. lithified) it is called tuff. Tephra are classified according to the size of the volcanic material; if it is smaller than 2 mm (0.08 in) in diameter it is called ash, if it is between 2 mm and 64 mm (2½ in) it is called lapilli. Larger particles (> 64 mm in diameter) are called volcanic bombs or blocks. Volcanic bombs are interesting and beautiful objects. They can be several metres in diameter and they usually have a round or ovoid shape due to deformation of the hot lava during the flight from the crater. Indeed, volcanic bombs are blobs of hot magma fired from an explosive volcano. During the flight, the surface quickly cools, acquiring a shiny, glassy appearance, but it also cracks due to cooling contraction. At the same time, the interior remains hot. The cooled surface further cracks on landing forming a special type of bomb known as the 'bread-crust bomb', which resembles a fresh loaf of bread. Bombs also spin during the flight acquiring a roundish or ovoid shape. On the other hand, volcanic ash has a small size and can be transported high in the troposphere and may travel far away from the volcano.

Deadly pyroclastic flows are composed of a mixture of glass shards, pumice, lapilli, loose crystals, rocks fragments, gases and ash (the major component). Upon cooling,

A volcanic bomb, with an ovoidal shape acquired during its flight from the vent. The bomb was erupted from Vesuvius.

the gases escape and the resulting rock is called ignimbrite. Pyroclastic flows can be relatively cold (<500–600°C or 932–1,112°F) and, in this case, the ignimbrite that forms is a relatively unconsolidated massive deposit. However, pyroclastic flows are often still hot when deposited, with still malleable pumice and ash due to the high temperature. This type of hot ignimbrite often collapses under its own weight and fragments fuse together forming a welded ignimbrite, while pumice will compact and flatten forming characteristic 'fiamme', after the Italian word for flame. Often the gas, still trapped in the main body of the ignimbrite, escapes forming fumarolic pipes or a gas-escape structure.

ABOVE: Ignimbrite with fiamme.

BELOW: Pumice (white) and scoria (black) from Stromboli, Italy. Pumice is the gas-rich froth of volcanic glass; it is very light and full of holes left by escaping gases. It is the only rock that floats in water.

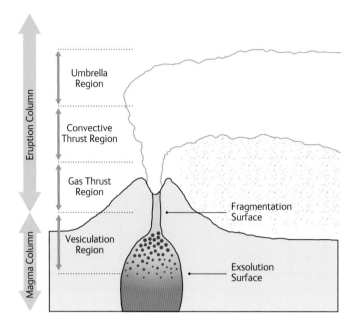

Eruption Column

Umbrella Region

Convective Thrust Region

Gas Thrust Region

Vesiculation Region

Magma Column

Fragmentation Surface

Exsolution Surface

Bubbles form in the magma in a process called vesiculation or gas exsolution. The bubbles grow and coalesce leaving the magma in the interstices between the bubbles. This process is called fragmentation because the magma is fragmented by the growing bubbles. At this point the explosive eruption will occur.

viscosity magma, like rhyolite, can be very explosive because it contains a lot of gas that cannot easily escape. The gas is trapped in the viscous magma but the bubbles continue to grow and expand until there are enough bubbles to fragment the magma and eject it explosively.

Types of volcanoes

Volcanoes come in different shapes and sizes, which reflect the type of magma they erupt and to a lesser extent the type of activity and where on Earth they occur. The different types of volcanoes are classified in terms of their morphological features and where they occur.

We have already encountered the stratovolcanoes, the giant cone-shaped hugely explosive volcanoes, emitting moderately viscous lava with intermediate to high silica contents. These volcanoes are found at convergent plate margins, the majority of them around the Ring of Fire. The most famous volcanoes of this type include Mt Fuji in Japan, Mt Rainier in the USA, Mt Vesuvius in Italy, Pinatubo in the Philippines, Merapi in Java, Indonesia and Colima and Popocatépetl in Mexico.

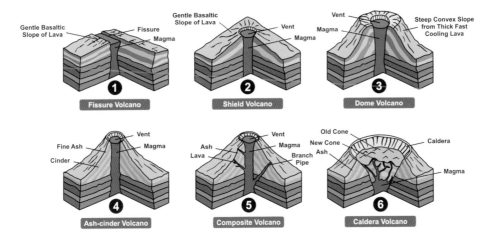

Volcanoes come in different shapes and sizes. The main types are illustrated here.

Puntiagudo volcano (on the left) and Osorno volcano (on the right) as seen from Casablanca volcano – three of the many stratovolcanoes in Northern Patagonia, Chile.

TOP: Cerro Azul volcano, a shield volcano, on Isabella Island, Galapagos Islands, Ecuador.
ABOVE: Popocatépetl volcano, a stratovolcano, in Mexico and the town of Cholula laying in its shadow.

Shield volcanoes, in which a central vent emits low-viscosity basaltic lava, are another common type of volcano, mostly found in intra-plate settings and associated with hot spots. The most famous examples are the Hawaiian volcanoes, and Erta Ale in Ethiopia, but shield volcanoes are also commonly found in the solar system, with many examples especially on Mars (e.g. Olympus Moon). The emitted lava has low viscosity so can travel long distances, producing a very large but low profile volcano resembling a warrior's shield. Despite their low profile, shield volcanoes can reach a considerable height as with Mauna Loa in Hawaii, which is 4,169 m (13,678 ft) in altitude. They are built by effusive and low explosive activity with lava flows and lava fountains. When the lava is very fluid and emitted along a fracture or a fissure, 'flood basalt' or 'plateau basalt' will form. Examples of this are the Columbia River basalt in the USA, the large Siberian Traps, the Parana Traps in Brazil and the Etendeka Traps in Namibia.

Very fluid lava, emitted along a fracture or fissure, forms flood basalt or plateau basalt as in the Columbia river basalt, USA.

Cinder cones are characteristic of Strombolian eruptions and other small to moderate eruptions with lava fountains. The most famous example is Paricutin volcano in Mexico. Lava domes are bulges of viscous silica-rich lava that commonly form in the craters of volcanoes and can be characteristics of many volcanoes. For example, Mt St Helens in the USA was characterized by the presence of a lava dome in its crater that erupted explosively in 1980.

Very viscous, silica-rich magma (i.e. rhyolite) produces big explosive eruptions, which can trigger the collapse of the volcano after the evacuation of magma leaving a circular or semi-circular depression, called a caldera. Caldera volcanoes can be several to tens of kilometres in size. They are rather flat volcanoes and for the gigantic ones such as the Campi Flegrei Field, Italy and Yellowstone, USA it is often hard to realize that you are actually walking inside the caldera of an extinct or dormant volcano. Gigantic calderas are often called supervolcanoes.

The crater of Colima volcano, Mexico, one of the most active volcanoes in North America. The activity is characterized by a dome building phase followed by vulcanian explosive eruptions that destroy the dome. This photo was taken during a survey flight in February 2015 and it is clear that there is no dome in the crater – it was destroyed by an explosive eruption a few months before.

TOP: Mt St Helens, USA. The amphitheatre shape is the result of the 1980 eruption, which caused the collapse of a portion of the volcano.

ABOVE: Cinder cone on the flank of the Popocatépetl volcano, Mexico.

Not only volcanoes, but supervolcanoes

Very large volcanoes, such as Yellowstone, have captured the attention of the media for their impressive destructive power. They have been labelled supervolcanoes, which despite not being a proper scientific definition is now widely accepted by the scientific community. Supervolcanoes are gigantic volcanoes able to produce extremely powerful eruptions with a magnitude 8 on the Volcanic Explosive Index (VEI) scale discharging more than 1,000 cubic km (240 cubic miles) of material. During such powerful eruptions, supervolcanoes also produce giant calderas, the most famous of which is Yellowstone, which is around 72 x 55 km (45 x 34 miles) in size. The most recent eruption occurred 630,000 years ago. In the last 2.1 million years, there have been at least another two supereruptions – one at 2.1 million years and the other 1.3 million years ago. However, a supervolcano is also characterized by smaller, 'normal' eruptions and at Yellowstone there have been many much smaller eruptions.

Yellowstone is not the only supervolcano on Earth. Indeed, there are several of them. Though the exact number is unknown, a rough estimate is around 20, including Long Valley in California, Valles Caldera in New Mexico, Toba in Indonesia, Taupo in New Zealand, Aira in Japan and Phlegraean Fields (Campi Flegrei) in Italy. The most recent supereruption is the Oruanui which occurred 26,000 years ago in New Zealand.

Some of the supervolcanoes are famous for the emission of hot water and steam (i.e. geothermal activity) or for their fumaroles emitting smelly gases. At Yellowstone there are about 10,000 geothermal features including geysers, hot springs, mud pots and fumaroles. The Campi Flegrei near Naples are famous for smelly fumaroles and the emission of sulfuric acid. All these activities are linked to the presence of a still hot and active magma body at depth, which is discharging volcanic gas and heating the groundwater.

It is quite hard to imagine what the impact of an eruption from Yellowstone would be, with possibly roughly 800 km (500 miles) around the volcano covered in ash. The nearby states of Montana, Idaho and Wyoming would be mostly directly affected but such a huge eruption would have a global impact, changing the climate for years. However, the chances of such a big eruption occurring are extremely low.

The closest we can get to imagine the effect of an eruption of a supervolcano is offered by the geological record of the Toba eruption. Some 73,000 years

The fumaroles of the supervolcano Campi Flegrei near Naples, Italy are famous for the emission of smelly gases and sulfuric acid.

ago the eruption of the Young Toba Tuff in northern Sumatra had such a huge impact on the climate and the environment, it is believed to have changed the course of our human ancestors, as we will see in Chapter 4. It was a gigantic eruption that expelled seven trillion tonnes of rocks, probably emitting about three gigatonnes of sulfur, although recent studies suggest a very low sulfur delivery in the atmosphere. The eruption contributed to the formation of the 100 x 30 km (62 x 18 miles) caldera that is currently occupied by Lake Toba. Indeed, Lake Toba is a nested caldera, the results of at least four volcanic eruptions each one producing a caldera overlapping with the previous one. The Young Toba Tuff eruption finally shaped the current depression occupied by the lake. Toba has not erupted historically but geophysical studies have identified a large magma chamber still present beneath the bottom of the lake. It has been calculated that at least 1% of the Earth's surface was covered by a 10 cm (4 in) blanket of ash dispersed by the immense volcanic plume of the Toba eruption. The Young Toba Tuff eruption had global consequences, although the effect on climate change is still debated. There is evidence that the Toba eruption occurred during a period of deteriorating climate conditions culminating in the last glaciation about 67,500 years ago. However, numerical modelling and paleoenvironmental evidence (ash beds identified in ice-cores and sediment cores from land and the seabed) indicate that the eruption triggered a drop in the average air temperature of around 2–3°C (36–37°F) (with up to 15–17°C) during a period that has been called a 'volcanic winter'. This cold period may have lasted some five to seven years. However, there is no evidence that the eruption initiated a glaciation.

Large Igneous Province (LIP) and mass extinction

Large Igneous Provinces or LIPs refer to gigantic plateaux covering areas of many thousands of square kilometres and up to several kilometres in thickness that occur both on continents and in oceans in the middle of a tectonic plate. They

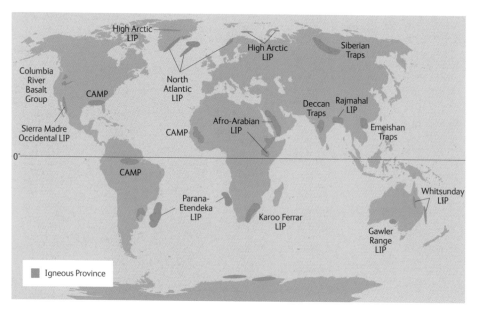

Worldwide distribution of the Large Igneous Provinces (LIP).

are mostly basaltic, but some silica-rich (rhyolitic) provinces are also known (e.g. Sierra Madre, Mexico; Chon Aike, South America). LIPs are best preserved in the Mesozoic and Cenozoic eras and they mostly comprise continental flood basalts, oceanic plateaux, volcanic rifted margins (where continental break-up is associated with magmatic activity) and a trail of volcanoes that stretch along the sea floor (called aseismic oceanic ridge). LIP is a loose term used to identify the outpouring of large amounts of prevalently basaltic magmas in a relatively short period of time of 1 to 5 million years (yes, this is short in geological time!). The processes that form LIPs are different from those active at a 'normal' oceanic ridge and convergent margins. In fact, LIPs are linked to an anomalous high thermal regime, which means a particularly high temperature of the Earth's mantle. They are also believed to be linked to mass extinctions.

The origin of LIPs is still a matter of debate and there is not a single theory that fits all the different LIPs. The different hypotheses all agree that there is a need for high thermal energy, able to produce large amounts of magma for a short period of time. How this high thermal energy is produced is not clear. The two most common, but not exhaustive, theories envisage the presence of a

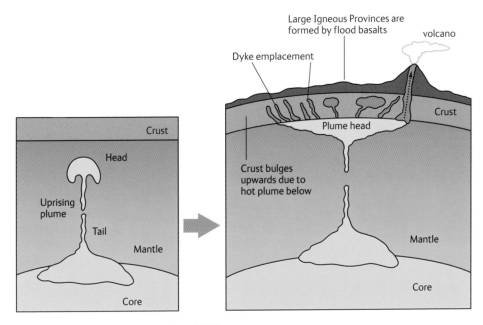

Schematic sketch showing the hypothesis of LIP formation by a mantle plume.

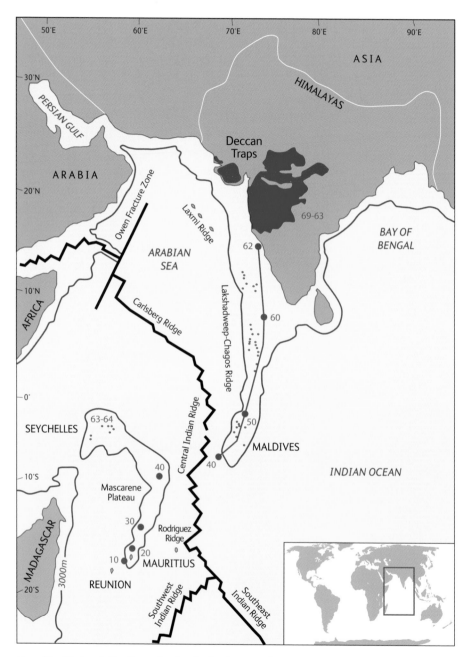

Map of the Western Indian Ocean showing the distribution and migration in time (red numbers are millions of years) of the LIP associated with the Réunion-Deccan plume.

mantle plume or a meteorite impact. Young mantle plumes can have a rounded head as large as 1,000 km (620 miles) in diameter and a narrow tail. Initial activity of a new mantle plume is expected to be associated with substantial uplift of the lithosphere of around 1,000 m (3,280 ft) and high temperatures causing extensive melting and high eruption rates in the plume head which corresponds with the centre of the volcanic province. The presence of a mantle plume can explain both the high thermal regime and high eruption rate, in the head of the plume, and the trail of volcanic islands associated with the tail of the plume such as those developing from the Mascarene Plateau toward Réunion in the Western Indian Ocean, or off the Deccan Traps into the Indian Ocean in Southern India.

On the other hand, the impact of a large meteor is capable of producing high energy in a localized yet ample area for a relatively short period of time. The idea that some LIPs can be triggered by a meteor impact was proposed to explain some characteristics of the rocks found in Canada at the Sudbury Complex (the so-called tachylite – a shock-induced glassy rock formed by fusion and subsequent rapid cooling of pre-existing rocks) and it is strengthened by the observation that some LIPs are contemporaneous with some impact events. However, the relationship between a meteor impact and the formation of LIP has yet to be firmly demonstrated, although in principle a large meteor impact could trigger shock-melting and subsequent decompression melting which, in turn, could produce an amount of magma comparable with the size of some LIPs. An impact-related origin has been proposed by some researchers for the largest mid-Cretaceous (about 120 million years ago) LIP, the Ontong Java Plateau, in the southwest Pacific; however this has been challenged by other researchers who point to a plume-related origin. The question is still open. It is worth keeping in mind that there is not a single theory that can explain all the different LIPs, each one has its own story.

The connection between LIPs and mass extinction is not firm either. However, the timing of LIPs and mass extinction strongly suggests some causal relationship. Indeed, four mass extinctions have taken place in the last 300 million years and they all coincide with LIP eruptions. However, not all LIP eruptions coincide with mass extinctions. It seems plausible that the eruption associated with LIP can release into the atmosphere large amounts of SO_2 and CO_2, which can have huge effects on climate, as we will explore in some detail in Chapter 4. In general, LIP events are associated with global warming rather than cooling and thus the

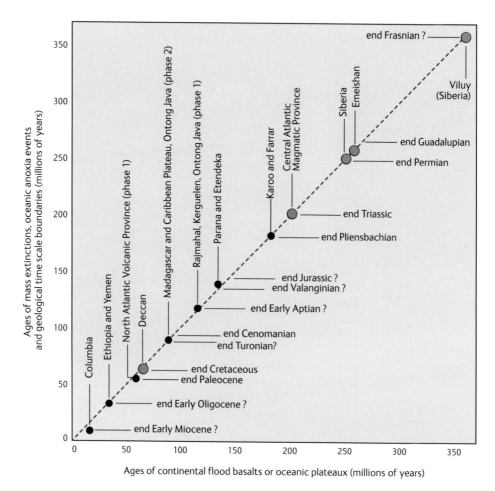

The connection between LIPs and mass extinction is not firm but the timing of LIPs and mass extinctions strongly suggests a causal relationship and four mass extinctions have taken place in the last 300 years that all coincide with LIP eruptions.

CO_2, an important heat-trapping (greenhouse) gas, can be responsible for that (SO_2 has a cooling effect instead). At the same time the CO_2 has also a deleterious effect on the ocean, inhibiting the life of carbonate-secreting organisms. However, several scientists note that the amount of CO_2 erupted during the lifetime of an LIP is not high enough to cause a sensible greenhouse effect and thus a modification in the average temperature. Another possibility is that these massive volcanic events can be associated with the emission of methane, which is an efficient greenhouse gas. Unfortunately, even in this case, the evidence is not solid enough and scientists continue to work hard to understand the causal effect between mass extinction and LIPs.

Volcanoes on other planets

Volcanoes are found not only on Earth, but they also occur on many other planets and asteroids. We have very good evidence of volcanic activity on Mars, Venus, the Moon, Jupiter's satellite Io, the asteroid Vesta and possibly Mercury. On Io images of volcanic plumes, similar to our terrestrial ones, have been captured by satellite. The surface of the Moon and Mars are dotted by calderas which are particularly gigantic on Mars. However, on none of these planetary bodies is there evidence of active plate tectonics as there is on Earth. Plate tectonics account for about 60% of the volcanism on Earth, along constructive and destructive margins (i.e. mid-ocean ridges and subduction zones, respectively). We still do not completely understand the reason why plate tectonics is present only on Earth. Many arguments point to the importance of extensive water bodies on the Earth's surface. These may favour subduction of wet marine sediment, which in turn can help generate the low-density continental crust encouraging the subduction. However, we still don't know what triggers plate tectonics and we have a long way to go before we have a complete understanding of our Earth and other planets. The best that we can do is to keep studying them.

3 Earthquakes and faults

The morning of Saturday, 1 November 1755 in Lisbon was unusually warm, as it had been during the previous days. The festivity of All Saints' Day and the clear sky encouraged many people to attend church early. A little after 9am the ground started to shake with vibrations from north to south, and with an oscillation of the buildings which, after the first minute of the quake, started to collapse. According to many people, the shaking lasted for six or seven minutes, with two pauses. There were narrow and long fractures along the ground and the dust from the collapsing buildings formed a fog that made it difficult to breath. Many survivors of the first shocks rushed to the docks where, some 40 minutes later, they observed the sea retreating. As the sea moved away from the shore it left the bottom uncovered, revealing shipwrecks. Soon afterwards, three giant waves engulfed the coast submerging many survivors and any buildings that remained. Once these waves had receded, fires broke out , enveloping the Palace of the Marquis of Louriçal and the Church of S. Domingo, the area around the Castle and many adjacent buildings. The fires continued for five days. In all, some 85% of the buildings of the town were destroyed and the death toll reached 30–40,000.

The earthquake caused extensive damage in the southern part of the Iberian Peninsula and Morocco, and was widely felt in southern Europe. The destruction of the Portuguese capital, with all its artistic treasures, including Vasco da Gama's travel records, raised a widespread commotion and started a lively debate about the causes of earthquakes among European scholars. The German philosopher, Immanuel Kant, wrote a textbook explaining that earthquakes were caused by the shifting of huge subterranean caverns filled with hot gases. The theory was not

OPPOSITE: The aftermath of the Great Kanto earthquake that struck Tokyo-Yokohama, Japan in September 1923, emanating from a seismic fault beneath the sea floor with the Philippine plate thrusting itself agaist the Eurasian plate.

new as it dated back to Greek and Roman philosophers, but Kant's book did much to disseminate that view. French philosophers also widely discussed the effect of the earthquake on the life of people – France being the centre of Illuminism at that time scholars were interested in the causes of natural phenomena and their effect on the everyday life of people.

Theories on the causes of earthquakes have changed substantially since these first proposals. Modern theory propounds that earthquakes result from the rupture of rocks that have been subjected for a long time to an accumulation of stress. This theory was formulated in 1910 by Henry Fielding Reid, Professor of Geology at Johns Hopkins University in the USA. The theory arose in the aftermath of the great San Francisco earthquake of 1906, and is called the theory of elastic rebound. According to this hypothesis, elastic energy is stored in the rocks that slowly deform, when subjected to stress arising from the movement of plates, until their internal strength is exceeded. Upon rupture and displacement of the rocks, the energy is released in the form of elastic waves which propagate into the Earth's interior. The displacement occurs along a surface called a fault.

A fault showing the displacement of rocks in a strata of volcanic ash at a roadside cutting in the Andes.

There are different types of fault according to the type of stress applied to the rocks. Normal faults derive from a tensile stress that tends to pull apart the rocks so that, upon rupture, there is a sliding of two blocks of rocks, and the distance between two points, laying on opposite sides of the fault, is increased. In a normal fault, the block above the fault (called the hanging wall) moves downwards relative to the block below the fault (the footwall). Reverse (thrust) faults occur when a compressive stress pushes the rocks, and the hanging wall above the fault moves upwards compared to the footwall. Strike slip faults result from a torque applied on the rocks so that after the rupture the two blocks of rock are displaced laterally.

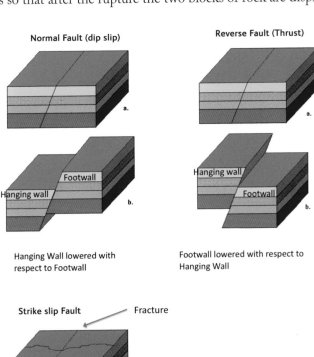

Normal Fault (dip slip)

a.

Footwall

Hanging wall

b.

Hanging Wall lowered with respect to Footwall

Reverse Fault (Thrust)

a.

Hanging wall

Footwall

b.

Footwall lowered with respect to Hanging Wall

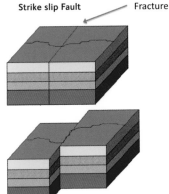

Strike slip Fault   Fracture

ABOVE: Normal faults result from stress that pulls aparts blocks of rocks so that the 'hanging wall' block moves downwards. A reverse fault occurs when the 'hanging wall' block moves upwards.

LEFT: In a strike slip fault the block of rock is displaced in a horizontal direction parallel to the line of the fault.

The motion of portions of the crust and mantle, called plates (see chapter 1), causes an accumulation of stress within the Earth. The relative motion of plates may cause compression, extension or lateral sliding, which may result in an earthquake along the borders of the plates. More commonly, each single plate does not move uniformly but is subjected to internal distortions that create an accumulation of stress within the plate or near its edges and, not always, the resulting faults involve the total length of the plate. As we will see later, the length of the fault is a measure of the magnitude of the earthquake.

Seismic waves

February 20th. This day has been memorable in the annals of Valdivia, for the most severe earthquake experienced by the oldest inhabitant. I happened to be on shore, and was lying down in the wood to rest myself. It came on suddenly, and lasted two minutes, but the time appeared much longer. The rocking of the ground was very sensible. The undulations appeared to my companion and myself to come from due east, whilst others thought they proceeded from south-west: this shows how difficult it sometimes is to perceive the directions of the vibrations. There was no difficulty in standing upright, but the motion made me almost giddy: it was something like the movement of a vessel in a little cross-ripple, or still more like that felt by a person skating over thin ice, which bends under the weight of his body. A bad earthquake at once destroys our oldest associations: the earth, the very emblem of solidity, has moved beneath our feet like a thin crust over a fluid; - one second of time has created in the mind a strange idea of insecurity, which hours of reflection would not have produced.

CHARLES DARWIN
Describing the earthquake of Concepción, Chile, 1835

The stored elastic energy propagates from the displaced fault (the source) in the form of elastic or seismic waves. The seismic waves transport energy with the oscillation of each elementary particle of mass, which is propagated to neighbouring particles. The modes of oscillation cause different types of elastic waves, the so-called body waves and the surface waves.

One type of body wave comprises an elastic oscillation of compression and expansion, (rarefaction, like in soundwaves) in which the particles oscillate back and forth along the direction of propagation. These are called P-waves (Primary) because they propagate with the fastest velocity. Transverse waves are body waves

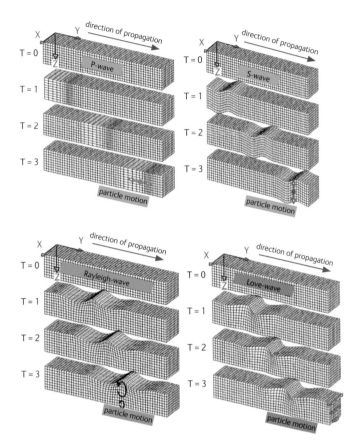

Different types of seismic waves showing how the direction of oscillation of the particles varies for each type of wave.

where the oscillation of particles is perpendicular to the direction of propagation. These are known as S-waves (Secondary) and are slower than P-waves.

Surface waves propagate on the Earth's surface and rapidly attenuate with depth. There are two kinds of surface waves: Rayleigh and Love. In Rayleigh waves, the particle motion occurs in the plane along the surface of the ground and in the direction of propagation. In the Love waves, the particle motion is transverse and parallel to the surface. Because of the different velocity of propagation, the seismic waves are recorded by a seismograph at successive times – first the P-wave, then the S-wave and finally the surface waves. The surface waves generally have a higher amplitude and last longer than body waves and, in large earthquakes, are the ones responsible for the damage to buildings.

THE SEISMOGRAPH

The seismograph or seismometer consists of an instrument with high density (the mass) suspended to a fixed support. As the seismic waves occur, the support moves with the soil, and the mass tends to remain steady, so there is a relative motion between the support and the mass. The suspended mass can oscillate either vertically or along a fixed horizontal direction, so that it is possible to separate the motion into three components. The first electromagnetic seismograph was invented by Giuseppe Palmieri, professor of Geophysics at the University of Naples and director of the Vesuvius Observatory, the first volcanology observatory in the world. This seismometer was able to detect the direction of the incoming seismic waves and record the arrival time. It was mainly utilized to study the seismicity associated with the eruption of Vesuvius. In older seismographs, the motion was recorded by means of a pen on the mass recording the displacement on a drum rotating with a constant velocity. In more recent seismographs the relative motion is digitally recorded.

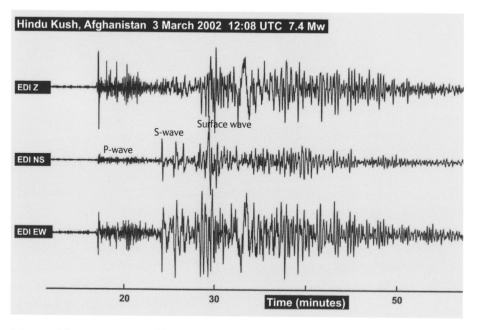

Seismograph from a seismogram in Edinburgh, Scotland of the earthquake of 3 March 2002, in the Hindu Kush mountains of Afghanistan, 15 minutes after it struck. Ground movement speed was recorded in three directions: vertical (top), north–south (middle) and east–west (bottom).

The location of an earthquake can be obtained by measuring the time of arrival of the different seismic waves on different seismometers. You need a minimum of three seismometers. Since the P- and S-waves travel with different velocity (S with a lower velocity), the difference in the arrival time of the two waves increases with distance. This difference represents a measure of the distance between the receiver (seismograph) and the source of the earthquake. The source is known as the hypocentre, and its projection on the Earth's surface is the epicentre. The circle with a centre on the seismograph and radius equal to the distance measured by the S–P time is the locus of all the points having the same distance from the earthquake. The intersection of three circles from three different seismographs gives the location of the earthquake.

More precise methods use the arrival time of P-waves and S-waves to more than three seismographs. Using this method, it is possible to identify the location of small earthquakes with an accuracy of hundreds of metres, even if

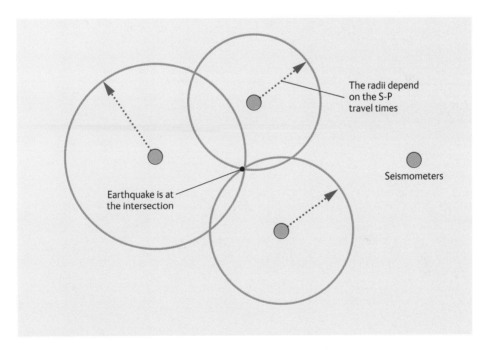

The radii depend on the S-P travel times

Seismometers

Earthquake is at the intersection

How to locate an earthquake using the S-P time, which gives the distance of the earthquake from the seismic station. The intersection of the three circles having the radius equal to the distance S-P provides the approximate location of the earthquake.

the actual source of the earthquake (the dimensions of the fault) may be tens of kilometres. The type of fault (normal, reverse or strike slip) can be identified by the distribution of the initial motion at different seismographs (either a pull or a push). The entire seismogram represents a record of the arrival of all different waves including those that have been reflected or refracted within the Earth.

Studying the Earth's interior

The seismic waves propagate with different velocities within the Earth, generally increasing with depth. Their velocity depends on the elastic properties – how much a body will deform under a given pressure – and the rigidity of the medium and its density.

The increase of velocity with depth is due to the change of elastic parameters of rocks. This property causes an upward bending, with depth, of the body waves that eventually emerge at the Earth's surface at a distance from the source. Seismometers located at increasing distances from the earthquake epicentre record the arrival of body waves that have travelled through increasing depths in

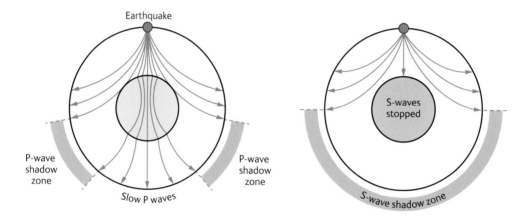

The propagation of seismic P- and S-waves and the effect of the discontinuity between the mantle and the outer core. The waves reaching the outer core are either reflected or refracted thus creating a shadow zone at the surface of the Earth, which varies according to the type of wave.

the Earth. Studies looking at the paths of seismic waves discovered that there is a shadow zone for the arrival of the direct P-waves (i.e. those waves that have not been reflected within the Earth) caused by an important discontinuity within the Earth – the transition between the mantle and the outer core. The S-waves do not propagate beyond this discontinuity, suggesting that the outer core is made of fluid matter. Further studies have discovered that the inner portion of the core is comprised of solid matter.

Several more studies have given a better picture about the transition between crust and mantle. In the upper portion of the mantle there is a slight reduction in the velocity of seismic waves (the low velocity zone). It is thought that this is associated with the presence of a certain fraction of liquid material, which may be the source of the generation of primary magmas.

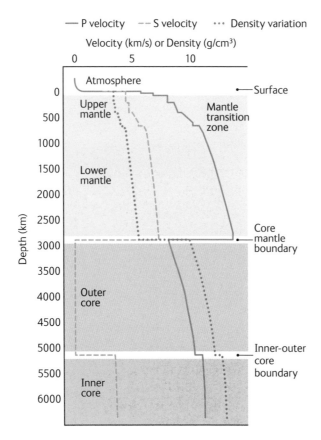

The variation of the velocity of propagation of seismic waves and the structure of the Earth.

Measuring an earthquake

The method of investigation which I purposed to adopt is based upon the very obvious truth, that the disturbances and dislocations of various solid objects by the shock of earthquake, if carefully observed with reference to their directions and extent of disturbance, and to the mechanical conditions in play, must afford the means of tracing back from these effects, the directions, velocities, and other circumstances of the movements or forces that caused them… In order, however, to make comparison, and to obtain some clear notion, of the relative distances of propagation of equal seismic effort, in different radii, from the seismic vertical, we must attempt to sketch out, the form and boundaries of equal effort, upon conventional principles. I have hence divided the total area, of the vast region within which the shock was in any way perceptible, into four, more or less concentric areas, each marked by a determined though arbitrary limit, of the seismic effort that acted within it.

ROBERT MALLET
Describing the Neapolitan earthquake, Italy, 1857

Robert Mallet, an Irish civil engineer, visited southern Italy after a great earthquake there in 1857 that damaged a wide area. He established the first principle of observational seismology and he drew lines of equal damage of the earthquake, called isoseisms, to better describe its effects. The extensive report sent to the Royal Society of London in 1862 marks the beginning of the quantitative observation of earthquakes.

OBSERVATIONAL MEASUREMENTS

In 1878 a more detailed scale, with the effects of earthquakes divided into 10 grades, was devised by two seismologists – the Italian Michele Stefano De Rossi and Francoise Alphonse Forel from Switzerland. This scale was modified by the Italian seismologist Giuseppe Mercalli in 1902, because it did not fully take into account the stronger earthquakes. Mercalli modified the scale further in 1908, adding two more grades following the suggestion of an Italian physicist Adolfo Cancani, in the aftermath of the great earthquake that destroyed the cities of Messina and Reggio Calabria in southern Italy. In 1930, the German geophysicist August Sieberg modified this scale once again, taking into account a wider classification of the effects of the earthquakes. Since then, the scale has been called the Mercalli–Cancani–Sieberg Scale or MCS scale. The scale was modified in 1931 by two American seismologists

Harry Wood and Frank Neumann, and later by Charles Richter in 1956, to adapt it to the types of American buildings, and was called Modified Mercalli (MM). Several other scales have been devised in more recent times to account for the modern characteristics of buildings. In Europe, the European Macroseismic Scale (EMS), reported in the table below, is currently used.

EMS intensity	Definition (description of typical observed effects)
I Not felt	Not felt.
II Scarcely felt	Felt only by very few individual people at rest in houses.
III Weak	Felt indoors by a few people. People at rest feel a swaying or light trembling.
IV Largely observed	Felt indoors by many people, outdoors by very few. A few people are awakened. Windows, doors and dishes rattle.
V Strong	Felt indoors by most, outdoors by few. Many sleeping people awake. A few are frightened. Buildings tremble throughout. Hanging objects swing considerably. Small objects are shifted. Doors and windows swing open or shut.
VI Slightly damaging	Many people are frightened and run outdoors. Some objects fall. Some houses suffer slight non-structural damage like hair-line cracks and small pieces of plaster may dislodge.
VII Damaging	Most people are frightened and run outdoors. Furniture is shifted and objects fall from shelves in large numbers. Many well-built ordinary buildings suffer moderate damage: small cracks in walls, fall of plaster, parts of chimneys fall down; older buildings may show large cracks in walls and failure of non-structural walls.
VIII Heavily damaging	Many people find it difficult to stand. Many houses have large cracks in walls. A few well-built ordinary buildings show serious failure of walls, while weak older structures may collapse.
IX Destructive	General panic. Many weak constructions collapse. Even well built ordinary buildings show very heavy damage: serious failure of walls and partial structural failure.
X Very destructive	Many ordinary well-built buildings collapse.
XI Devastating	Most ordinary well-built buildings collapse, even some with good earthquake-resistant design are destroyed.
XII Completely devastating	Almost all buildings are destroyed.

The European Macroseismic Scale (EMS) reflects twelve different observational effects for increasing earthquake intensity.

The Mercalli scale, and those derived from it, measure the strength of an earthquake through its effects on humans and buildings, but it does not indicate the energy of the earthquake. So a strong earthquake, occurring in a desert zone, cannot be graded by such a scale.

PHYSICAL METHODS OF MEASUREMENT

The ability to classify earthquakes through physical methods arose through the development of seismometers. In 1935 the seismologist Charles Richter started to classify the earthquakes of southern California using a method based on a specific seismometer, the Wood-Anderson, at a given amplification. The method was devised to establish empirically the relationship between the maximum seismographic amplitudes of given shocks at various distances. According to this method, the magnitude was measured by the logarithm of the maximum amplitude (measured in mm) corrected with the distance. This method started to be used, firstly in the USA and then successively all over the world and became known as the Richter Scale. Within the scale, the different intensities of earthquake are graded by means of their magnitude. The Richter magnitude is also called the local magnitude because it cannot be applied to earthquakes at a distance greater than 600–700 (373–435 miles) away. It is written as M_L where L stands for local. Other types of magnitude have been used to classify earthquakes, indicating the amplitude of the body wave (m_b) or the surface wave (M_S). There are, however, some shortcomings when measuring larger earthquakes. The Richter magnitude underestimates the earthquakes with magnitude above 7 and there are other problems with the use of m_b and M_S.

In order to estimate the Richter magnitude of an earthquake the amplitude is directly measured on the seismogram. The S–P time is necessary in order to estimate the distance for the correction of the magnitude. In 1979, the American Thomas Hanks and the Japanese-American Hiroo Kanamori tried to overcome this problem by relating the magnitude (M_w) to the size of the fault, and using a rupture measure called seismic moment (M_0). A better characterization of the largest earthquakes has been obtained using this magnitude. The largest known earthquake occurred in Chile in 1960 and had been previously classified with an M_S= 8.3 using the M_S scale.

Year	Name	M$_w$
1952	Russia: Kamchatka Peninsula	9.0
1960	Chile: Puerto Montt, Valdivia	9.5
1964	Alaska	9.2
2004	Indonesia: Sumatra: Aceh: Off West Coast	9.1
2011	Japan: Honshu	9.1

Table showing the largest earthquakes that have happend on Earth since 1900.

The calculation of the Moment magnitude (M_w) requires the analysis of the entire spectrum of the signal and takes some time to be evaluated. This means that the first evaluations are often based on the M_S scale. The use of the different scales often creates problems with the press and media that do not understand the difference between the various evaluations.

SEISMIC RISK

We have just had news, my esteemed Lucilius, that Pompeii, the celebrated city in Campania, has been overwhelmed in an earthquake, which shook all the surrounding districts as well. The city, you know, lies on a beautiful bay, running far back from the open sea, and is surrounded by two converging shores, on the one side that of Surrentum and Stabiae, on the other that of Herculaneum. The disaster happened in winter, a period for which our forefathers used to claim immunity from such dangers. On the 5th of February, in the consulship of Regulus and Virginius, this shock occurred, involving widespread destruction over the whole province of Campania; the district had never been without risk of such a calamity, but had been hitherto exempt from it, having escaped time after time from groundless alarm.

The extent of the disaster may be gathered from a few details. Part of the town of Herculaneum fell; the buildings left standing are very insecure. The colony of Nuceria had painful experiences of the shock, but sustained no damage. Naples was just touched by what might have proved a great disaster to it; many private houses suffered, but no public building was destroyed.

SENECA

In *Naturales Quaestiones*, 65 AD

The philosopher Seneca recalls, with the words on the previous page, the moderate earthquake that occurred in 62 AD, which destroyed the city of Pompeii only 17 years before the volcanic eruption of 79 AD. Earthquakes have always affected human life. An earthquake in 464 BC caused extensive damage in Sparta and triggered a slave revolt, straining the tension with Athens. A violent earthquake in 17 AD, caused extensive damage to 13 cities in Lydia (now western Turkey) and is recalled by the historian Tacitus. To commemorate the event and pay for the damages, Emperor Tiberius struck a coin with the inscription 'CIVITATIBVS ASIAE RESTITVTIS' (Asian Cities Rebuilt).

The effect of earthquakes may be the total destruction of buildings because the shaking of the ground causes a violent acceleration, whose value may be larger than the gravity acceleration (9.8 m/s^2). The collapse of buildings is caused mostly by the tangential forces of the surface waves which have a large amplitude. One of the most damaging earthquakes of our times occurred in China. On 28 July 1976, a magnitude M_W=7.3 struck the city of Tangshan located at about 150 km (93 miles) to the east of Beijing. The earthquake destroyed 95% of the buildings in the town and the death toll was 242,769 people because it struck an area with a high density of population and

The destruction caused by the 1976 earthquake in Tangshan, China.

The liquefaction of soil during the 1964 earthquake of Niigata, Japan caused
the entire buildings to fall without damage to the buildings themselves.

poorly constructed buildings. An important step in understanding the damage
caused by an earthquake was made after the violent earthquake in Ischia, Italy in
1883. This earthquake (with a magnitude probably lower than M_L=5) destroyed
the town of Casamicciola on Ischia Island in southern Italy. It was noted that
some buildings around the village square remained unharmed, while others
close by were completely destroyed. This observation motivated the first research
into the response of different types of ground to seismic waves as, for example,
the liquefaction of the ground, which may loses its internal strength because of
the vibration with the result that the ground does not support the overlaying
buildings.

The occurrence of earthquakes is governed by the dynamics of plate tectonics
and is more violent near the edges of convergent plates. Their distribution over
time is controlled by the accumulation of stress within the lithosphere and the
same faults may generate similar earthquakes over a time period of millennia.

Although it is presently impossible to forecast the occurrence of an earthquake,
it is possible to draw a map of the areas where there is a higher probability of

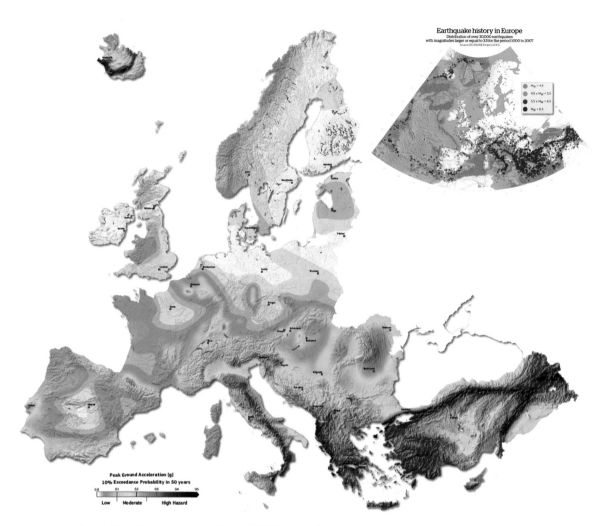

Seismic Hazard Harmonization in Europe (SHARE) map produced to
provide a seismic hazard model for the Euro-Mediterranean region.

earthquakes, based on the distribution of known faults and the past seismic
history of the area. In Europe several scientific institutes, under the sponsorship
of the Seventh Framework Program of the European Commission, have
launched a program called Seismic Hazard Harmonization in Europe (SHARE)
whose main objective is to provide a community-based seismic hazard model for
the Euro-Mediterranean region. One of the first results of the programme was

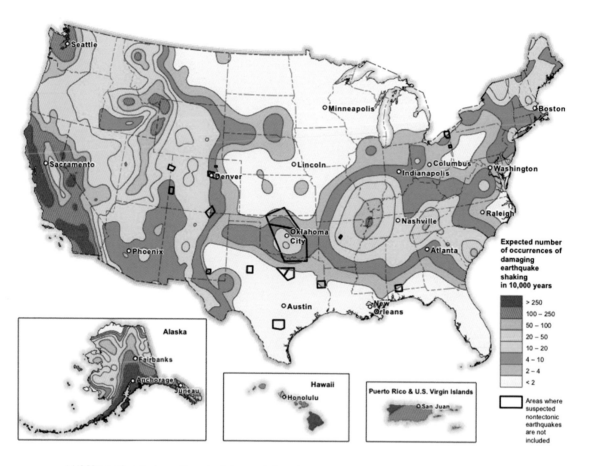

Expected number of occurrences of damaging earthquake shaking in 10,000 years

> 250

100 – 250

50 – 100

20 – 50

10 – 20

4 – 10

2 – 4

< 2

Areas where suspected nontectonic earthquakes are not included

A USGS map showing how often scientists expect damaging earthquake shaking around the USA, used to inform the public about the hazard from earthquake ground shaking in their area.

the compilation of a Probabilistic Seismic Hazard map of Europe and Turkey. The map shows where ground shaking (i.e. maximum horizontal ground acceleration) would be reached or exceeded, with a 10% probability, in 50 years. This means that the average recurrence of such ground motions occurs every 475 years, as prescribed by the national building codes for standard buildings in Europe. The darker coloured areas in Italy, Greece and northern Turkey, correspond to peak ground acceleration of the order of 0.5 times the gravity acceleration (4.9 m/s^2), which could cause the collapse of all buildings.

Tsunami (and earthquakes)

That morning I was playing soccer with my friends. We ran home after the strong earthquake and after that I heard a really loud noise, like an aeroplane. When I looked at the sea I saw something I had never seen before and I was terrified. My family rushed into our minivan but the road was full with everyone trying to escape. The black wave hit our minivan, turning us over several times before I blacked out. When I regained consciousness, I was in the water. Holding on to a school chair, I floated until I landed on a beach. I had no idea where I was and I was so hungry and thirsty. There were bodies and debris everywhere. Under a mangrove tree I saw a mattress had washed up and I started searching for packets of noodles and bottles of water, collecting them around my mattress. After five days I didn't have any water or food left. I survived there by myself until day 20. That's when I saw people coming to collect the bodies. They rescued me and took me to Fakinah hospital where I found my father. He told me my mother and sister had died in the tsunami.

TESTIMONY OF MR MARTHUNIS, 17
From Alue Naga village, Banda Aceh, Indonesia, 2004

Events since the turn of the century have tragically made the general public aware of a hazard that was previously known only to specialists and the communities living on the shores of the Pacific. The images, spread worldwide by the media, of the events following the two major earthquakes of this century – that of Sumatra in 2004 and of Fukushima, Japan, in 2011 – showed giant sea waves penetrating for hundreds of metres into the land, destroying everything in their path and causing the death of several hundred thousand people. The term tsunami is Japanese and literally means 'great harbour wave'. It is observed only near coastlines. A tsunami is a sea wave with a period (the distance between wave crests) that is extremely long – from one hundred to several hundred kilometres. In open water it may pass unnoticed but with a wave height of several tens of metres, it is able to overcome any barrier when it reaches the coastline. Its destructive force is made worse when the water retreats, because it transports with it all the dangerous debris caused by the impact of the violent wave. The first sign of a tsunami is the retreat of the sea, which may leave the sea floor uncovered for hundreds of metres. The retreat is immediately followed by the arrival of the first wave, which moves onto the land like a sudden rising of the seawater, a fast-moving tide. A tsunami that has been caused by the most violent earthquakes may have multiple waves and last for hours.

Satellite image of the town of Soma, Fukushima prefecture, Japan, taken before the 2011 earthquake (ABOVE) and one day after (ABOVE RIGHT), showing the effects of the resulting tsunami.

Tsunamis are not only generated by earthquakes. Submarine landslides or volcanic eruptions can cause them too. The association with earthquakes is complex, but in general they are the result of violent thrust earthquakes that occur at shallow depths in the plate below the sea. The effect of the thrust is a sudden displacement of a large volume of water, with the subsequent propagation of the sea disturbance from the source. The tsunami waves travel outwards at high velocity (increasing with the depth of the sea) all over the ocean and may hit areas where the generating earthquake has not been felt. The tsunami generated by a magnitude $Ms=7.1$ earthquake in the Aleutian Islands in 1946, devastated the islands of Hawaii and caused 159 deaths and $US26 million in damages. After this catastrophe, seismologists in Alaska established the first warning centre for tsunamis. The warning network was further enhanced after the 1964 Alaska earthquake, with the creation in 1967 of the Palmer Observatory in Palmer, Alaska. This was built under the auspices of the Coast and Geodetic Survey of the United States and in 2013 it became the National Tsunami Warning Center (NTWC), responsible for issuing warnings of tsunami that may affect the US coastline. After the 2004 Indian Ocean earthquake and tsunami catastrophe the Pacific Tsunami Warning Center in Hawaii was delegated with issuing warnings for the Asian part of the Pacific Ocean as well.

4 Impacts and benefits

Qui su l'arida schiena
Del formidabil monte
Sterminator Vesevo,
La qual null'altro allegra arbor nè fiore,
Tuoi cespi solitari intorno spargi,
Odorata ginestra,
Contenta dei deserti.....

Questi campi cosparsi
Di ceneri infeconde, e ricoperti
Dell'impietrata lava,
Che sotto i passi al peregrin risona....

Fragrant broom,
content with deserts:
here on the arid slope of Vesuvius,
that formidable mountain, the destroyer...

...These fields scattered
with barren ash, covered
with solid lava,
that resounds under the traveller's feet...

G. LEOPARDI
Broom, or the flower of the desert, 1836

According to the US Geological Survey (USGS), there are about 1,500 potentially active volcanoes worldwide and about 500 of them have erupted in historical time. This number does not consider the volcanoes on the ocean floor (https://www.usgs.gov). On any single day as many as 10–20 volcanoes can be actively erupting somewhere on Earth. As I write, there are 18 currently erupting volcanoes reported by the Global Volcanism Program of the Smithsonian Institution, National Museum of Natural History (http://volcano.si.edu). Over 800 million people live within 100 km (62 miles) of an active volcano, but why do people live in the vicinity of a volcano despite the obvious danger?

In this chapter, we explore the impact of volcanoes on our environment and on our lives – the dangers as well as the benefits for those who live in their proximity and for those living very far away. Volcanoes impact our life in so many and unexpected ways, from mineral resources, to fertile soil, geothermal energy, even our cats benefit from volcanoes in their litter! They have influenced art and our climate, but they also pose a threat to the people living in their proximity. There are different types of volcanic risk and scientists are working towards a better understanding of the complex mechanisms behind volcanic eruptions in order to mitigate those risks, as we will see in the next chapter.

OPPOSITE: The umbrella-like volcanic plume, generated by the Plinian eruption of Mount Pinatubo, Philippines in 1991, rising up into the atmosphere.

Volcanoes and climate

On 15 June 1991, Pinatubo volcano in the Philippines awoke with the second-largest volcanic eruption of the 20th century. The eruption produced a cloud of volcanic ash, spreading over hundreds of kilometres, giant pyroclastic flows and mudflows. The impact of the eruption was huge with long-term effects on both the population and the climate. Fortunately, the indigenous population of Aeta highlanders had been relocated from their homes on the slope of the volcano before the eruption, because of precursory signals of an eruption in the months beforehand. Some 18 years later, most of them are still living in resettlement camps. The 200,000 people living in the lowlands around Pinatubo face continuous threats of lahars (a mudflow or debris flow formed by loose pyroclastic materials remobilized by rain from the slope of a volcano), that can bury their town at any time during typhoons. The eruption also had a global effect on climate. Indeed, volcanic eruptions inject gases into the atmosphere, particularly sulfur, which is converted into sulfate aerosols. The aerosols cool the Earth's surface by scattering some solar radiation back, but, at the same time, heat the stratosphere by absorbing both solar and terrestrial radiation The combined effect can produce enhanced polar-to-equator temperature gradients, that can have an important impact on the climate. In fact, an eruption in tropical latitudes can produce winter warming and summer cooling in the northern hemisphere continents of up to 1°C (34°F), but mostly 0.3–0.8°C (32.5–33.4°F).

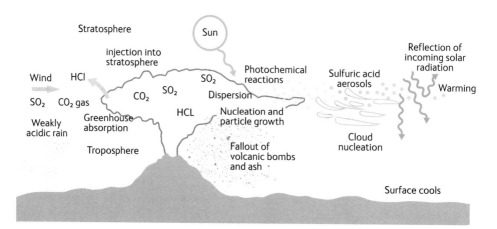

Volcanic inputs into the atmosphere and their effects.

It took 22 days for the Pinatubo volcanic cloud to complete the circumnavigation of the globe. It has been calculated that three megatonnes of sulfur dioxide disappeared in the first week. The sulfur was transformed into a mixture of sulfuric acid and water, in the form of minute particles that stayed in the stratosphere for months causing optical effects and scattering solar radiation. A total of about 20 million tons of sulfur dioxide were injected into the stratosphere and were dispersed around the globe causing a drop in the air temperature of 0.5°C (33°F), during the period 1991–1993. The Pinatubo eruption offered scientists the opportunity to study the ongoing effect of a volcanic eruption. Subsequently, scientists started to look at past eruptions in search of evidence of how eruptions affected the climate.

The 1815 eruption of Tambora, a volcano in the Sumbawa Islands of Indonesia, had an even bigger global effect. The eruption occurred in April 1815 and was the largest eruption of the past 500 years with a magnitude 7 on the Volcanic Explosive Index (VEI). The volcano ejected 160 km³ (38½ cu miles) of erupted material and the volcanic plume reached an altitude of about 43 km (27 miles), while the explosion was heard 2,000 km (1,243 miles) away. When a large part of the volcano collapsed, it triggered a tsunami with a wave of up to 4 km (2½ miles). In total the eruption killed 60,000 people. In Europe, nobody paid attention to the very sparse and late news of the eruption, which was mentioned by newspapers only in November, 7 months after the eruption (the telegraph had not been invented at the time). However, that distant and unknown eruption was responsible, as we now understand, for crop failure, famine and unusual weather during the miserable 1816, known as the 'year without summer'. It is unknown how many people suffered from the aftermath of the eruption in Europe, Asia, eastern USA and Canada, but the record of unusually cold and rainy weather, especially during the summer, is clear. The consequences were devastating, particularly regarding crop failure, which then induced a severe famine. There is a claim that the defeat of Emperor Napoleon Bonaparte in the Battle of Waterloo (18 June 1815) by the British–Prussian Coalition, led by the Duke of Wellington, was in part due to the eruption of Tambora. The extremely and unusually wet weather made the battlefield a pool of mud, which delayed the starting of the battle, allowing the union of Prussian and Anglo–Dutch forces. As Victor Hugo put it in *Les Misérables*: 'Had it not rained on the night of 17th/18th June 1815, the future of Europe would have been different... an unseasonably clouded sky sufficed to bring about the collapse of a World.' It is an unproven claim, but it is possible that a distant volcano might have played a big part in European, and human, history. As we will see in many other examples later on, this was not the first time.

Another powerful eruption that changed the world was the eruption of Krakatau volcano (or Krakatoa) in Indonesia on 26 and 27 August 1883. Many scientific papers and science-based fictional works have been written about this eruption and the main reason is because it was the first huge volcanic eruption to make the headlines around the world, thanks to newly installed submarine cables that carried telegraphic communications. It was also the first major eruption studied in detail, with a complete report published by the Royal Society of London in 1888. The optical effects of volcanic particles travelling in the atmosphere were striking, producing a haze in shades of pink, lilac and purple. The colourful skies lasted many months and were captured in many paintings and especially those of JMW Turner. Much has been written about the violent detonations on the day of the eruption. The strongest one occurred on the morning of 27 August 1883 and was heard across about 10% of the Earth at a distance of up to 4,800 km (2,983 miles).

> It was at the very moment when Schruit was watching the Loudon steaming toward the safety of port that there came the first roar of the explosion. It was an extraordinary sound, he thought: far, far louder than anything he recalled from before. He looked sharply over to his left and saw, instantly, the unforgettable sight of a tremendous eruption.
>
> SIMON WINCHESTER
> *Krakatoa. the day the world exploded,* 2004

The eruption of Krakatau was one of the largest of the last 200 years but it was neither exceptional nor the most intense in terms of explosivity, rating 6 on the VEI. Despite that, it was truly a catastrophe. Half of the mountain disappeared during the eruption displacing a huge amount of seawater and triggering a major tsunami that inundated the shorelines of Java and Sumatra, affecting tens of thousands of people. An indirect consequence of the eruption of Krakatau is that it contributed, together with many more direct events, to the collapse of the colonial society in Sumatra and Java. The Royal Society of London reported that rafts of pumice had been observed floating on water alongside the remains of human corpses. The pumice rafts were seen drifting in the Indian Ocean and as far as the shoreline of Africa.

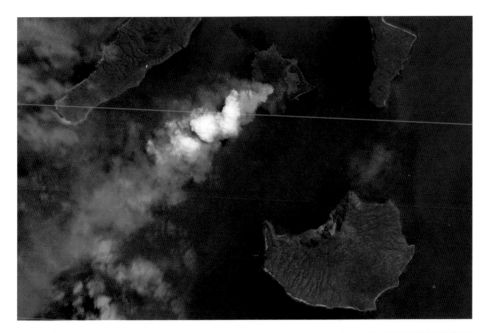

Anak Krakatau, Indonesia as seen in September 2018 from satellites, with volcanic ash streaming southwest over the Sunda Strait (TOP) and from the Strait itself (ABOVE). The volcano appeared in 1927 and has been very active since then.

Pumice, a very light and porous volcanic rock formed when a gas-rich froth of glassy lava solidifies rapidly. They are mostly produced during Plinian eruptions.

'…about the third week in July 1884, the boys… were much amused by finding on the beach stones who would float, evidently pumice stones. The lady who was with them… also noticed that there were a quantity of human skulls and bones… quite clean with no flesh remaining on them…'

'The British ship Bay of Naples… had reported… when 120 miles from Java's Frist Point, during the volcanic disturbance, she had encountered carcasses of animals even those of tigers, and about 150 human corpses… besides enormous trunks of tree borne along by the current.'

SIMON WINCHESTER,
Krakatoa. the day the world exploded, 2004

Krakatau started to grow again and by 1927 a new island called Anak Krakatau (child of Krakatau) emerged only to disappear and re-emerge several times. Today, what we see is the island that reappeared in 1931 and has grown since then and the volcano is still very active. The most recent eruption, so far, occurred in December 2018. It also triggered a tsunami, smaller than the 1883 one, which nevertheless caused death and destruction.

Volcanoes: the environment and resources

Volcanoes play an extremely important role on our planet. Our atmosphere is the result of volcanic activity. Early Earth's atmosphere 4.5–3.5 billion years ago was dominated by carbon dioxide (CO_2) released by volcanoes and this carbon dioxide was fundamental for photosynthesis. Buried organic carbon, and the photosynthetic process, led to the accumulation of oxygen and volcanic vapour (H_2O) condensing to form the oceans. It took 2 billion years to reach a significant concentration of oxygen and since then the atmosphere has evolved to its present composition.

Volcanoes have also been fundamental to the origin of life via submarine hydrothermal vents. Known as submarine 'smokers', they are heated by volcanic activity. The temperature can vary from hot environments (up to 400°C/752°F) to a warm one (50–90°C/122–194°F). Generally the hotter vents are called black smokers because they emit clouds of black particles rich in sulfur-bearing minerals; these minerals are characteristic of the deep-sea environment. The colder vents mostly emit white clouds of materials and they are called white smokers. Submarine smokers can provide a warm reactive habitat suitable for hosting a variety of microbial communities. On dry land, solfatara fields, consisting of soil, mud holes and surface water heated by volcanoes (up to 100°C/212°F at the surface, higher at depth), represent another environment in which microbial life flourishes. Such hyperthermophiles, believed to be among the first living cells, need only water, trace minerals and heat. All these ingredients can be found at both hydrothermal vents and solfatara fields. In the following pages some of the main resources linked to volcanic environment will be explored.

GEOTHERMAL ENERGY

An intrusion of magma into the crust is a strong source of heat for both surrounding rocks and groundwater. If the surrounding rocks are permeable and there is enough groundwater available, a geothermal system may form, which can be harnessed to generate electricity. The first geothermal field to be harnessed in that way was that of Larderello in central Italy in 1904. It is still active and one of the biggest in the world, producing about 10% of the world's supply of geothermal energy. In Iceland, up to 25% of the country's total electricity comes from geothermal sources, with about nine out of 10 houses heated with it. Geothermal energy is a sustainable form of energy whose production can be prolonged by optimizing the methods used to generate it.

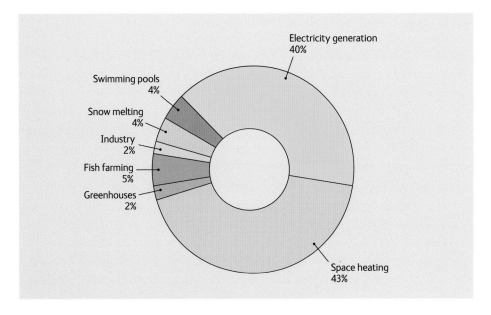

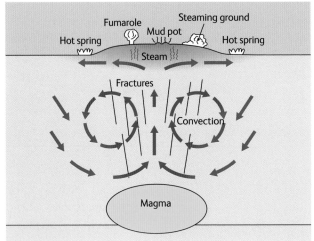

TOP: Diagram showing how the geothermal energy generated in Iceland was used in 2013.

ABOVE: Relationship between the geothermal system and magmatic activity. The blue arrows show the circulation of meteoric water that recharges the system and gets heated by the heat coming from the magma. The heated water moves upwards (red arrows) delivering heat to shallow levels where boiling geothermal water and steam can reach the surface.

Hot water and steam are also used for bathing and for food preparation, especially by early humans. The black eggs (called Kuro-Tamago) of Owakudani, a geothermally active area in Japan, are a local specialty. Chicken eggs are boiled for 60 minutes in natural hot spring water at 80°C (176°F) and then steamed at 100°C (212°F) for about 15 minutes. The sulfur and iron of the thermal water and volcanic gas turn the eggs black with a deliciously volcanic taste. It is believed that eating the eggs is very healthy, potentially adding a few years to your life.

MINERAL RESOURCES

We are dependent on mineral resources. We use them in almost every aspect of our daily lives from the components in computers, cars and cellphones to fertilizers, detergents and even the litter used in cat trays. Mineral resources is a loose term covering a variety of materials (such as ores, building stones and fossil fuels, etc) of intrinsic economic value (i.e. there is enough material to allow economic extraction). Mineral resources have been formed in a variety of geological processes including magmatic and hydrothermal environments.

Ore deposits are a variety of mineral resource: a mass of rock that contains useful elements (mostly metals such as copper, gold and iron among the most important, but also non-metals like fluorite and asbestos) in a concentration and quantity sufficiently high enough to be mined economically. Many types of ore deposits are generated in volcano- (or magma-) related environments. Volcanic ore deposits include both metallic ores (such as porphyry copper, Au-Ag and Zn-Cu-Pb deposits) and non-metallic ores (such as gypsum and phosphate rocks used as fertilizer) but also industrial minerals (including clays and zeolites).

There is little doubt that the metallic ores (copper, nickel, zinc, gold, silver, platinum group elements and rare earth elements) hosted by volcanic rocks are essential to our technologies. The exploitation of such ores is a driving force of economic development and, at the same time, a societal challenge with a potentially dramatic impact on humans and the environment. It is beyond the scope of this book to discuss the effect of the mining process on the environment and on the population living in mining areas, but it is sufficient to say that there is a strong debate between the scientific community and mining companies. It is important to stress that there is a growing attention to reconcile mining activity with environmental and local population protection without losing economic profit.

Sulfur is one of the most abundant volcanic gases both in the form of sulfur dioxide and hydrogen sulfide. It solidifies in yellow lumps around the crater of active volcanoes. It is mined in many active volcanoes and, once processed, it is used for sugar bleaching, matches, fireworks, pesticides, medical treatment for skin diseases, fertilizers, vulcanized rubber, glass production, cement mixtures and the manufacture of adhesives.

In East Java, Indonesia, sulfur represents the life and death of the many men who mine it by hand from the active crater of Ijen volcano with absolutely no protection. The miners, carrying up to 90 kg (2 lb) of sulfur, climb for 200 m (656 ft) from the crater to the top of the crater rim and then down 3 km (1.9 miles) to the outer slope. This journey is made several times a day. It is a very dangerous job. Poisonous volcanic gases, such as hydrogen sulfide and sulfur dioxide, burn the workers' eyes and throats; given time it can even dissolve teeth. This traditional method of sulfur mining was used in the past in many countries, including Italy, Chile and New Zealand. Today, modern methods allow the production of sulfur in a less deadly way – as a by-product resulting from other industrial processes, such as oil refining.

Sulfur lumps (yellow) in a volcanic lava. Sulfur lumps solidify from hydrogen sulfide gas and sulfur dioxide around the crater of active volcanoes.

Sulfur miners in East Java, Indonesia climb into the crater to retrieve lumps of sulfur at the sulfur mine near the Kawah Ijen volcano, on the island of Java, Indonesia.

Volcanoes and humans

The relationship between volcanoes and humans is very strong, dating back to our ancestors and to the still highly debated human origins. There are strong arguments supporting the importance of volcanic activity in influencing human evolution, its adaptation and migration, with the East African rift valley playing a key role. Pre-human apes and the first hominids walking on two feet roamed the East African rift. Their footprints are preserved in tephra about 3.6 million years old in the area of Laetoli in Tanzania. *Homo habilis* appeared there around 2.3 million years ago followed, at around 1.8 million years, by *Homo erectus*, an ancestor of modern humans. With a changing diet from vegetarian to omnivore, they had to adapt from forest to savannah and face the danger of fierce predators. The unique geological activity of the East African rift valley, with continuous rejuvenation of the rift by volcanic eruptions and earthquakes, might have provided the ideal environments for our ancestors. It has been suggested that the challenge of a continuously changing and very rough environment, such as that created by intense volcanic activity, provided protection, at the same time stimulating evolution and adaptability.

Some of the earliest evidence attributed to *Homo sapiens*, our direct ancestors, was found in Ethiopia dating back to about 195,000 years ago, where they experienced periods of colder conditions, changing climate and cataclysmic caldera eruptions. Modern genetic methods indicate relatively small numbers for these early humans compared to present-day populations, but also a complex origin from different parts of Africa for our species. By 100,000 years ago, modern humans were beginning to spread out from Africa across southern Asia, eventually using boats to reach remote regions like Australia, more than 60,000 years ago.

Around 74,0000 years ago there was the gigantic eruption (magnitude M_e 8.8, where e is the magnitude of the eruption based on the total mass of erupted materials) of Toba supervolcano in northern Sumatra, creating a caldera 100 km (60 miles) long by 30 km (19 miles) wide, that is today occupied by Lake Toba (see also Chapter 2). The eruption is known as the Younger Toba Tuff. Toba has not erupted historically, but the region is still volcanically and tectonically active. Indeed, the southern rim of the caldera is located on a fracture zone related to the Great Sumatra Fault, in turn associated with the subduction of the Indian plate beneath the Sunda plate and the megathrust that produced the giant 2004 Boxing Day Sumatra-Andaman earthquake and tsunami.

A combined effort of volcanologists and paleoanthropologists, using DNA and other data, has led to the claim that the Young Toba Tuff eruption 74,000 years ago almost wiped out our ancestors due to the severe volcanic winter that followed. It is claimed that the human population decreased dramatically to only a few thousands, possibly as a consequence of the extensive famine associated with a decade of winter, followed by a further thousand years where the climate was cold and dry.

Intriguing as it might be, this theory is far from accepted and counter-arguments include the fact that evidence of a volcanic winter associated with the Young Toba Tuff eruption is far from convincing. On the contrary, studies of Greenland ice cores indicate that the sulfur emission associated with the eruption, which would have caused the significant climate change, was much smaller than previously thought. At the same time, climate cooling was well underway before the eruption and the resolution of the palaeoclimate record and climate models are insufficient to relate it exactly to the Toba eruption. In addition, no other species, except humans, show a reduced population and low diversity of mitochondrial DNA. This is a strong argument, since the effect of such a large eruption, with an extended period of volcanic winter, would have impacted living species globally. Finally, the fossil record of Asian mammals doesn't support extinction, not even in the Toba area.

All in all, it seems unlikely that we are the descendants of a small group that survived a gigantic eruption. However, we know that volcanoes can significantly affect our lives in both positive and negative ways, as discussed above. There are many examples of famous eruptions that had a wide impact on human life or, as Clive Oppenheimer, a volcanologist from Cambridge University, puts it 'that shook the world'. Lots has been written about the Pompeii eruption in 79 AD and the human casts as well as the impact of the eruption of Santorini volcano in Greece, during the Minoan period in the Late Bronze Age (1500–1700 years BC), and of earthquakes on the Mycenean civilizations. One less well-known story is of the town of Tetimpa and the eruption of Popocatépetl volcano in Mexico.

TETIMPA

Popocatépetl volcano in Mexico is one of the most active volcanoes on Earth and it also ranks very high in terms of threatened population, with more than 20 million people living within 70 km (43 miles) of the crater. Continuously active since 1994, the present-day edifice has been shaped by voluminous lava flows

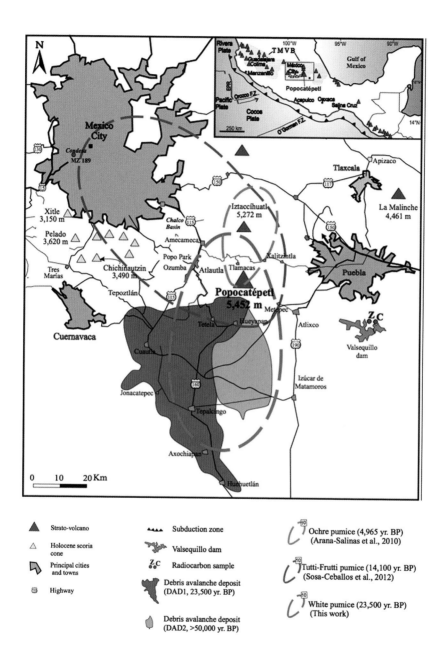

Map of Popocatépetl volcano with major nearby cities and towns. The
10 cm (4 in) isopachs are shown for three of the major Plinian eruptions
that have happened in the last 23,000 years.

punctuated by at least five major Plinian eruptions in the last 23,000 years. The Pumice with Andesite Plinian eruption occurred around 14,000 years ago and was the most powerful (VEI 6) on record at Popocatépetl, with ash fall reaching as far as Mexico City.

In the first century AD (about 2,000 years ago), another large VEI 6 Plinian eruption took place, producing an eruptive column that rose 20–30 km (12½ –18½ miles) before depositing an estimated minimum of 3.2 km³ (0.77 cu miles) of yellow andesitic pumice over more than 240 km² (93 sq miles) in an area extending at least 25 km (15½ miles) east of the volcano's crater. Pumice and ash completely buried the village of Tetimpa, a large dispersed farming village on the northeastern flank of the volcano, preserving it forever. At the same time the collapsing eruptive column produced devastating pyroclastic flows which buried the northwestern flank towards the Basin of Mexico.

Popocatépetl volcano, Mexico, one of the most active volcanoes in North America.

Pumices and ash layers from the Plinian eruptions at
Popocatépetl volcano, Mexico.

Volcanologists believe that this Plinian eruption took place during the dry
season, from October through May, when the prevailing winds blow west to east,
explaining the dispersion of the tephra towards the east. Archaeological evidence
supports this interpretation, indicating that the agricultural fields were not planted
at the time and there was a shortage of supply in domestic storage facilities.

The short- and long-term effects of the eruption were devastating. Short-term
effects included the loss of seeds for the next season, while long-term impacts
included the destruction of hunting areas and agricultural soil. Thousands
of people were prompted to relocate, moving north towards Teotihuacan and
further east towards Cholula. With time Teotihuacan saw a huge population
increase, also aided by another devastating eruption of Popocatépetl around

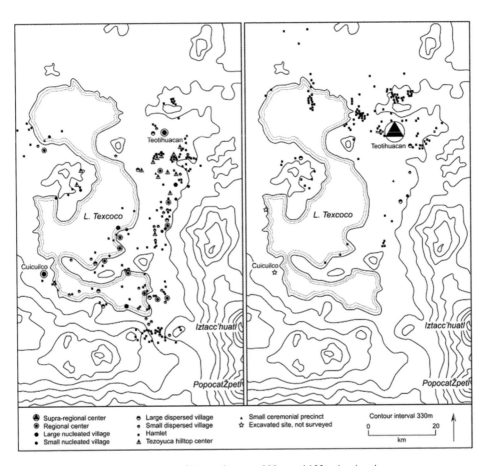

Distribution of settlements in the Basin of Mexico between 300 BC and 100 BC (**LEFT**) and 100 BC and 100 AD (**RIGHT**) and 100 AD showing the stronger concentration of populations around Teotihuacan with time as people moved north away from the volcano.–

1,000 years ago, becoming the most powerful city of pre-Hispanic Mexico. Today Popocatépetl, known by the local population as Don Goyo, is still very active, with daily explosions mostly confined to the crater, but frequent larger explosions with ash reaching as far as Puebla (about 64 km or 40 miles to the east) and Mexico City (about 90 km or 56 miles to the northwest). From time to time volcanic ash from Popocatépetl reaches the international airport of Mexico City, causing problems to the civil aviation and to the millions of people living nearby. In the very moment I am writing, Don Goyo had one of these relatively stronger eruptions whose detonation was clearly heard in Puebla.

VOLCANOES AND ART

Fuji no yama nomi ga chausu no ooi kana/cha-usu

Mount Fuji
like the tea-grinding mill
carried by the lice…

MATSUO BASHO, AGE 33,
1644–1694

Mt Fuji in Japan is perhaps the most famous volcano in the world. The first painted images date back to the Heian Period (794–1185). It is called Fujisan, san as a suffix is used to denote a sign of respect, and is celebrated widely in poetry and art, most famously by Katsushika Hokusaki's *One Hundred Views of Mt Fuji*. Katsushika Hokusaki was a celebrated Japanese artist of the Edo Period who lived between 1760 and 1849. Vesuvius is another iconic volcano highly celebrated in poems, prose and art.

Red Fuji Southern Wind, a famous painting by Katsushika Hokusaki.

Directly or indirectly, volcanoes have influenced the arts in various ways. Mary Shelley's *Frankenstein* and Lord Byron's *Darkness* would probably have never been written, if both artists hadn't been forced to stay indoors during their holidays during the miserable summer of 1816. As discussed earlier, Tambora volcano erupted in 1815 in Indonesia, but it affected the European climate a year later and particularly during the summer 1816. The eruption of Krakatau in 1883 also inspired great artworks – notably Turner's famous depiction of the pink, lilac and purple tint of the sunset over the British sky.

The power and unpredictability of volcanoes have inspired legends and superstitions. Many rituals are performed even today to seek the benevolence of the volcano, which may be humanized (Popocatépetl is called Don Goyo, short for Gregory from St Gregory), deified (Hawaii is believed to be the home of the Hawaiian goddess Pele) or highly respected (Fuji-san). One famous ritual is the traditional Neapolitan feast of Saint Gennaro's blood miracle. It is believed that if the saint's blood, contained in an ampulla and usually solid, liquifies on 19 September every year, it is going to be a good year for Naples, which also implies Vesuvius is not going to erupt. Residents of towns near Popocatépetl in Mexico have their own annual ritual on 12 March. Popocatépetl means 'the smoking mountain' in Nahuatl language and every year the locals climb the volcano where they prepare altars adorned with flowers, bowls of fruit and tequila. These gifts are left as offerings to Don Goyo to keep him 'smoking' happily.

⑤ Volcanic hazards

A cloud, from which mountain was uncertain, at this distance (but it was found afterwards to come from Mount Vesuvius), was ascending, the appearance of which I cannot give you a more exact description of than by likening it to that of a pine tree, for it shot up to a great height in the form of a very tall trunk, which spread itself out at the top into a sort of branches; occasioned, I imagine, either by a sudden gust of air that impelled it, the force of which decreased as it advanced upwards, or the cloud itself being pressed back again by its own weight, expanded in the manner I have mentioned; it appeared sometimes bright and sometimes dark and spotted, according as it was either more or less impregnated with earth and cinders.

PLINY THE YOUNGER
Letter LXV to Cornelius Tacitus

Volcanic hazards

There are many types of volcanic hazard related to the type of volcanic activity and the specific characteristics of each volcano. It is possible to define the 'hazards footprint' as the area likely to be affected by particular hazards. The concept covers the exposure of the area to a variety of natural hazards, along with the physical characteristics of the area itself. There are many volcanic phenomena that pose a threat, including ash falls, tephra and ballistic (e.g. volcanic bombs, rock falls) fallout, pyroclastic flows, gas emissions, lava flows, lahars, debris avalanches, landslides, and tsunamis. Volcanic hazards can be roughly divided into two main types: primary/direct and secondary/indirect. The direct hazards are linked to the volcanic eruption itself, such as fallout (ash, tephra and ballistic), flows (pyroclastic and lava) and gas emissions. Indirect hazards include all the

OPPOSITE: Mt Merapi, Java, a typical stratovolcano hugely explosive and with a giant cone, emitting moderately viscous lava.

events that do not necessarily occur during a volcanic eruption, but are related to the existence of a volcano, its morphology and the type of rocks emitted (such as loose rocks and steep flank). Lahars, debris avalanches and landslides are all secondary effects of volcanic activity and therefore part of the second type of volcanic hazard. Tsunamis are a special case since they can be both directly related to a volcanic eruption or they can be triggered by a subsequent landslide, as well as by earthquakes.

In this chapter, the main types of volcanic hazards will be explored along with examples and mitigation strategies. Volcano monitoring and eruption forecasting will also be analyzed alongside the information derived from a forensic volcanology approach – studying the rock record and the eruptive history of a volcano.

Volcanic ash

'… then again we were immersed in thick darkness, and a heavy shower of ashes rained upon us, which we were obliged every now and then to stand up to shake off, otherwise we should have been crushed and buried in the heap…'

'… At last this dreadful darkness was dissipated by degrees, like a cloud or smoke; the real day returned, and even the sun shone out, though with a lurid light, like when an eclipse is coming on. Every object that presented itself to our eyes (which were extremely weakened) seemed changed, being covered deep with ashes as if with snow.'

PLINY THE YOUNGER
Letter LXVI to Cornelius Tacitus

Volcanic ash is produced during explosive eruptions and can be transported great distances by the wind and the circulation of the high atmosphere. The ash emitted during the 1991 Pinatubo eruption circumnavigated the globe in 22 days and the eruptive column reached 34 km (21 miles) in height. Ash is extremely hazardous – it causes respiratory problems, eye inflammations, clogs air-conditioning units and water systems, disrupts critical infrastructure, causes roofs to collapse, and poses problems to the aviation circulation even in localities relatively far from the erupting volcano. In 2010 the eruption of the Eyjafjallajökull volcano in Iceland, caused the sudden and protracted (up to five weeks) closure of large portions of European airspace with huge economic loss for civil aviation. Volcanic ash can cover soil, plants and edifices, and it is also very difficult to remove.

Volcanic ash fall from Eyjafjallajökull volcano, Iceland, 2010, which caused the closure of large portions of European airspace.

The distance the ash can travel strongly depends on wind conditions and the amount of ash emitted, therefore not only the type and size of eruption is important but also the geographic location of the volcano and the time of the year. Ash is dispersed by the wind and its distribution is controlled by wind direction and the deposit thins out with increasing distance from the volcano. Between June to October there is a higher risk that ash from Popocatépetl will reach Mexico City since during this period winds blow westwards over Mexico City, whereas from November to May ash is more likely to reach the city of Puebla towards the east.

Ash is the finest portion of tephra, a loose term that indicates rocks produced during an explosive eruption when the magma is fragmented by the expanding volcanic gases (see pp.56–57). Larger tephra (bombs and blocks) do not travel far – less than 5 km (3 miles), follow ballistic trajectories and pose a local threat. Pumices are lighter and they can be transported for greater distances along with the ash, although their dimension will decrease with distance from the volcano. The city of Pompeii was covered by 3–4 m (10–13 ft) of pumice fallout during the initial phase of the 79 AD eruption of Vesuvius. Ash fall, however, is the most frequent and widespread volcanic hazard, because ash is also emitted during small to moderate explosive eruptions. Its impact can be extensive and last for a long period of time, because the eruption, or the 'eruptive period', can last from days to years. In addition, the ash may be remobilized and re-deposited by wind, rain (i.e. lahar) or human-activity for a long period of time after the eruption. Most of the casualties during the 1991 eruption of Pinatubo were the result of roofs collapsing under the weight of about 15 cm (6 in) of wet ash. Some 320 people were killed and 279 injured. Wet ash is heavier than dry ash and caused the collapse of the local flat roofs. Volcanic ash also infiltrates into cracks, causes electrical short circuits and affects mechanical systems. Near the volcano, it can trigger intense lightning storms due to the electrical charge of interacting particles in the volcanic plume.

One important aspect of the ash hazard is the effect on agriculture and animal stock. Indeed, as with humans, ash can cause respiratory and eye inflammation to animals as well. It is not clear how animals are affected by eating grass contaminated by ash, but this might be particularly important in areas near a constantly erupting volcano. Fluorosis, dentition and bone damage and respiratory

Cars covered in ash during the eruption of Pinatubo volcano, Philippines, 1991. Most casualties resulted from roofs collapsing under the wet ash.

problems might be common in such areas. Finally, ash is very difficult to clean and remove. In contrast to heavy snow, which can be removed from road, roofs and cities, accumulated somewhere and after few weeks will melt, ash doesn't melt. It needs to be manually removed and disposed of at high economic cost, adding a further high cost during the aftermath of a volcanic eruption.

Lava flows

Lava flows are very rarely deadly events – there have been 824 casualties since 1900. However, they can be very disruptive since they can burn, crush and bury and are unstoppable, although in some cases they have been successfully diverted. In the very moment I am writing (July 2018) Kilauea volcano in Hawaii has been erupting for over two months. Kīlauea is located on the southeastern flank of the much bigger Mauna Loa volcano, and for many years it was thought to be a mere satellite of Mauna Loa, not a separate volcano. Hawaii is a volcanic island whose volcanic activity is linked to the presence of a mantle plume or hot spot (see chapter 2, Volcanoes). It is the youngest island of the Hawaii-Emperor volcanic chain. Its volcanic activity is very different from the big stratovolcanoes (such as Pinatubo, Tambora, Krakatoa, Popocatépetl and Fuji) that we have seen so far. It is a shield volcano characterized by basaltic lava flows, which are fluid and can flow for many kilometers, and has limited explosive activity, mostly as fire fountaining (see chapter 2, Volcanoes).

The 2018 eruption at Kilauea started in the afternoon of 3 May, after a period of increased seismicity that began on 30 April. The shallow summit of Kilauea caldera, Halema`uma`u crater, was occupied by a lava lake which started to drain from a fissure on the eastern flank, the so-called Eastern Rift Zone, toward the urban area of the Leilani Estates. Eventually the lava lake completely disappeared and the floor of the caldera subsided at a rate of 6–8 cm (2–3 in) per day, widening the caldera and causing frequent earthquakes. The lava flow, emitted by up to 22 fissures, has reached the sea, partially destroying the urbanized area of Leilani Estates. A great source of very accurate information and spectacular videos is the Hawaii Volcano Observatory (USGS-HVO) website. No causalities are reported, but many people had to be relocated. They lost their houses and had their lives completely torn apart with major economic damages. At the same time, the landscape around Kilauea has completely changed and rejuvenated.

Lava flow from Kilauea volcano, Hawaii, June 2018. The lava is
outpouring from one of the many fissures active during the eruption.

The lava flow from Kilauea was slow enough to allow people to escape, but in
the case of Nyiragongo volcano, in the Democratic Republic of Congo, the lava
moves at velocities up to 60 km/h (55 miles/h). This is due to the lava's unusual high
temperature (about 1,370°C, 743°F) and composition, with low in SO_2 and rich in
CO_2. During the eruption of Nyiragongo in 1977 fast-moving lava swept through
several villages killing an estimated 600 people. Nyiragongo is responsible for about
92% of casualties related to lava flows since 1900. The later eruption, on 17 January
2002, was even more devastating as lava flows poured from the lava lake occupying
the Nyiragongo crater, at 3,470 m (11,385 ft) above sea level, rapidly reaching the city
of Goma 17 km (10½ miiles) away on the shore of Lake Kivu. It has been estimated
that 25 million m³ (883 million ft³) of lava erupted from vents and several fractures.

Lava lake with a lava stream in Nyiragongo volcano, Democratic Republic of Congo. The lava lake is characterized by very fluid lava which moves at high velocity. It is one of the most dangerous volcanoes in the world.

The city of Goma was affected by two lava flows. About 13% of Goma was destroyed and an estimated 21% of the electricity network was damaged, alongside most of the airport and most of the economic resources. The eruption caused the death of about 150 people, mostly by CO_2 asphyxiation and from the explosion of a petrol station, and 470 were injured. The scale of the disaster forced the evacuation of 300,000–400,000 people towards the border of neighbouring Rwanda. This was truly an exodus which escalated the humanitarian tragedy that was already taking place at the border between the Democratic Republic of Congo and Rwanda due to ongoing ethnic and military conflict.

Heimaey town, Iceland (TOP) as seen from the Eldfell volcano. A house submerged by scoriae produced during the 1973 eruption (ABOVE) of the volcano just visible at the top of the image.

Scientists define mitigation as the act of reducing the impact of a natural hazard, while alleviating the loss. In the case of lava flows, the best mitigation strategy is to be prepared by planning for losses at an individual, local, regional and national level. Some attempts to deviate the lava flow via barriers, or by explosively destroying the wall of a lava tube to force the lava in a different direction, have been successful in Italy, Japan, Iceland and Hawaii. During the 1973 eruption of Eldfell volcano on the island of Heimaey in Iceland, the Icelanders sprayed a large amount of seawater over the lava flows to slow its movement in the attempt to save the town of Vestmannaeyjar and its port. The huge effort was successful overall although part of the town was destroyed. After the eruption, the residents returned to rebuild their town and make good use of the geothermal energy. Even the best of such tactics are controversial and costly projects with a limited lifespan. They can only buy time and can never stop the eruption.

Pyroclastic flows

On 3 June 2018, the Fuego volcano in Guatemala erupted, killing at least 113 people and leaving 332 missing. Thousands of victims were displaced and are still living in temporary shelters. People at Fuego were killed by a pyroclastic flow, the deadliest of the volcanic hazards. Pyroclastic flows, or pyroclastic density currents, are very hot avalanches of volcanic ash, pumice, gases and rocks, produced by the collapse of the explosive eruptive columns under their own weight (see chapter 2, Volcanoes). Pyroclastic flows reach temperatures of 400–500°C (752–932°F) and move at extremely high speed – greater than 300 km/h (186 miles/h) – down the flank of the volcano, wiping out everything in their path. They can travel farther than 100 km (62 miles) and cover an area larger than 20,000 km² (7,722 sq miles). They leave behind total destruction and hot, thick pyroclastic deposits that may be several 1,000 km³ (240 cu miles) in volume, called ignimbrite when rich in pumice. Nothing survives on their path and they cannot be outrun. The only way to survive is by not being there.

There are many examples of deadly pyroclastic flows. The most famous one is that associated with the destruction of Herculaneum and Pompeii by the eruption of Vesuvius in 79 AD, as we have seen in chapter 2, Volcanoes. Pyroclastic flows are commonly associated with the collapse of the Plinian eruptive column, but they can be produced by a lateral blast associated with the explosion of a gas-rich lava dome, as was the case with the 1980 eruption

Pyroclastic flow at Mount Sinabung, Indonesia during the January 2014 eruption. Pyroclastic flows are hot avalanches of volcanic ash, pumices, gases and rocks that move at high speed; they are the deadliest of the volcanic hazards.

of Mt St Helens and the 1902 eruption of Mt Pelée in Martinique. In the case of the 1991 eruption of Unzen volcano in Japan the pyroclastic flow was generated by the gravitational collapse of the lava dome. Both Mt St Helens and Unzen claimed the lives of volcanologists (Katia and Maurice Krafft and Harry Glicken at Unzen, and David A. Johnston at Mt St Helens) together with other people, and completely destroyed the area affected by the event. The eruption of Mt Pelée on 8 May 1902 completely destroyed the town of St Pierre,

Martinique killing its entire population of about 28,000 people, apart from one prisoner locked in the dungeon of the town's jail.

The landscape left after a pyroclastic flow is appalling: trees are flattened, houses are damaged and burned, and everything is buried under several metres of pyroclastic deposit. The main hazards are the lethal combination of high temperature, fine particles in the air, dynamic pressure, i.e. the pressure of the moving materials, high density and high speed. The flow picks up everything (loose rocks, trees, construction material, etc) encountered along its path, contributing further to the force of any impact. The effect of high temperature

The 8 May 1980 eruption of Mt St Helens, USA was devastating. The lateral blast caused a rockslide, augmented by the pyroclastic flow produced by the exploding dome, and the resulting devasting debris avalanches destroyed an area of hundreds of square kilometres flattening all the trees in the surrounding forests. Fifty-seven people were killed along with thousands of animals. The snow, ice and glaciers on the volcano melted, creating lahars which travelled for more than 80 km (50 miles) and caused further devastation.

is clearly seen in many victims of the Pompeii eruption, but also in the upsetting footage from the recent eruption at Fuego in Guatemala. Victims assume a 'pugilistic pose', very characteristic of death under high temperature. Remobilized dense material (e.g. loose rocks, trees, construction material) captured by the pyroclastic density current has the same force and effect as a missile, enhancing the damaging effects of the pyroclastic flow. A stunning example has been unearthed very recently at Pompeii. Archaeologists found the remains of a man decapitated by a stone, maybe a doorjamb, while trying to flee the fury of the pyroclastic flow during the 79 AD eruption.

ABOVE: Cast of an adult with a child, both victims of the 79 AD Pompeii eruption of Mt Vesuvius, Italy. The pugilistic pose with contracted hands is typical of death under the strain of high temperature.

OPPOSITE: Night eruption with lava fountaining and lava flow at Fuego volcano, Guatemala.

SOME USEFUL TERMINOLOGY

Pyroclastic density current – hot gravity-driven current composed of ash and gas that moves laterally along the ground. It is a general term that encompasses both the dense pyroclastic flows and the more diluted surges.

Ignimbrite – rock formed by the solidification of a pyroclastic density current deposit composed of pumice, ash and rock fragments (lithics), usually formed during large explosive eruptions. Can be welded or non-welded.

Lateral blast – a rapid decompression of a lava dome on a volcano due to a sudden collapse that creates a pyroclastic flow, which is ejected horizontally instead of vertically.

Block-and-ash flow – small volume of pyroclastic density current deposit prevalently composed of dense coarse blocks of rock fragments, generated by the collapse of a lava dome, in a medium to coarse ash matrix.

Volcanic unrest – period in which the volcano deviates from the background behaviour showing changes in one or more of the following parameters: seismicity (number, location and types of earthquakes), surface deformation, gas emission, geochemistry of fumarolic activity. A period of volcanic unrest might or might not lead to an eruption.

Lahars

On 2 May 2008 a large Plinian eruption began on the small (1,122 m/3,681 ft above sea level) and rather unknown Chaitén volcano in the southern Chilean Andes. The eruption started abruptly, with very limited precursory seismic signals 36 hours before the eruption. It was detected by instruments located more than 300 km (186 miles) from the volcano. The Plinian eruptive column reached 20 km (12.4 miles) in altitude and lasted for about six hours and it was associated with minor pyroclastic flows. It was followed by a second similar Plinian column on 6 May and a third one two days later. Between 10–12 May a new dome started to grow inside the crater accompanied by small ash columns and steam emissions. This activity continued until October. The ash plume led to the closure of regional airports with hundreds of flights between Chile and Argentina cancelled. The ash also heavily affected the aquaculture industry in the nearby Corcovado Gulf. However, the real disaster struck on 12 May in the form of floods and lahars – rain-triggered mudflows – which cut Chaitén town in two, completely destroying half of it. Luckily enough the 4,625 residents were already evacuated after the first Plinian event on 2 May, but the economic losses were extensive (estimated to be around US$12 million) and so was the impact on the population. Some 10 years on from this catastrophic event, the people of Chaitén are still coming to

TOP: House destroyed by the lahar – a rain-triggered mud flow – which occured a few days after the eruption of Chaiten volcano, Chile in 2008.

ABOVE: The volcanic dome as it appears inside the crater of Chaiten volcano, Chile, in 2016, 8 years after the eruption that partly destroyed the town of Chaiten.

terms with their losses. But humans are very resilient and there is now a project, led by the Fundación ProCultura, to transform the destroyed area of the town into an open-air volcano museum, which will preserve the legacy of Chaitén and showcase what has been done by local population and scientists since then to prevent a similar disaster in the future.

Lahars are mudflows produced by rain that remobilizes the pyroclastic material on the flanks of the volcano, producing a type of debris flow. There is no need for an eruption to trigger a lahar, in fact they can occur in volcanic areas even decades after the eruption has occurred. This is what happened on 5–6 May 1998 at Sarno and Quindici, two towns in southern Italy, when a series of lahars, triggered by intense rainfall, killed more than 150 people. Sarno and Quindici are located in Campania, at the foot of some rugged and steep reliefs about 30 km (19 miles) east of Vesuvius.

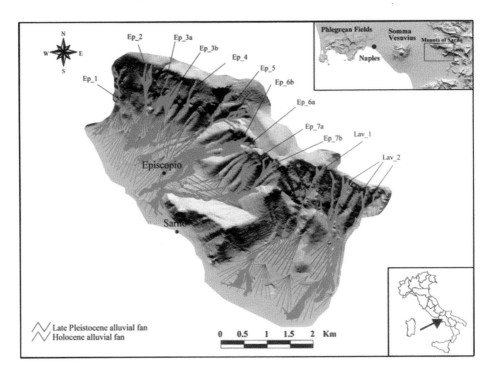

Digital elevation model (DEM) of the Sarno area (about 30 km, 19 miles, east of Vesuvius, Italy) showing, in green, the areas devastated by the 5–6 May 1998 lahar when intense rain remobilized loose materials, mostly produced by previous eruptions of Vesuvius, from the flank of the reliefs surrounding the area. Red and blue lines indicate alluvial fans.

This area is downwind from Vesuvius so it is covered by pyroclastic material from the volcano's past eruptions. That year exceptional rainfall, aided by the destruction (by agricultural activities) over the years of the vegetation covering the steep slopes, made a deadly combination, so that lahars were formed from about 34 small drainage basins.

The Sarno example is an extreme case of a volcano-related hazard due to unusual circumstances and to careless environmental maintenance. However, deadly lahars are quite common in volcanic areas during or immediately after an eruption when the fresh, loose material is easily remobilized by heavy rainfall. The town of Armero in Colombia was buried by a lahar following the melting of the glacier-cap of Nevado del Ruiz volcano during the 1985 eruption. The lahar killed 22,000 people.

Forecasting volcanic hazards and risk reduction

One of the main goals of volcanology is to be able to forecast the next eruption in order to save lives, protect assets and increase resilience in communities. Volcanic eruptions, unlike earthquakes, are typically preceded by warning activity, such as volcanic seismicity, deformation of the volcanic edifices or an increase in gas emissions from fumaroles. The precursory signals can start days or months in advance. A recent study found that about 50% of stratovolcanoes erupted within one month of the reported unrest (see box, p.126). However, there are many cases where the period of volcanic unrest doesn't culminate in an eruption. The activity can sometimes prompt false alarms, which are problematic since they reduce the trust of local communities.

Detection and recognition of early warning signals is the best way to anticipate, plan and mitigate an impeding volcanic eruption. To this end we need to know our volcano very well and keep a constant eye on it. Monitoring is fundamental, but it is also essential to have an in-depth knowledge of the past behaviour of the volcano. This allows volcanologists to develop a conceptual model of how the volcano works, particularly in the less visible portion of it –the plumbing system, where the magma is located and the volcano is 'brewing' the eruption. Monitoring data are essential for short-term forecasting, but longer-term forecasting is more effectively done by combining monitoring data with a forensic approach to volcanology. The latter uses geological, petrological (i.e. genesis and composition of rocks) and geochronological (i.e. rock dating) data to produce a conceptual model of the volcano that can be used to interpret monitoring data, and can help to quantify uncertainty, enabling risk reduction and management.

VOLCANO MONITORING

Systematic monitoring of volcanoes began in 1845 at Osservatorio Vesuviano on Vesuvius volcano in Naples, Italy. Luigi Palmieri, director of the Osservatorio between 1855 and 1872, invented several instruments that allowed the detection of volcanic signals for the first time. Between 1911 and 1914 the director of the Osservatorio Vesuviano was Giuseppe Mercalli, who defined the Mercalli Scale to measure the intensity of earthquakes as well as classifying volcanic eruptions. After the establishment of the Osservatorio Vesuviano, the science of volcano monitoring was born, leading to the foundation of volcano observatories worldwide. Dedicated volcano observatories now undertake volcano monitoring, provide volcanic hazards information and, in most cases, set up volcano alert levels, and issue forecasts of future activity. The alert is usually in the form of a colour code: green, yellow and red, with each colour reflecting a different level of concern, red being the highest alert when the eruption is imminent or occurring. Volcano observatories might also work in connection with the local civil defense organization.

The early warning signs of a volcanic eruption can be monitored by instruments both in-situ and remotely. This is mostly done by volcano observatories. Monitoring occurs at ground level, from the air and even from space. Data collected over a long time period helps to build the baseline of the normal behaviour of a volcano, so that a period of unrest can be more easily recognized at an early stage. Most volcanoes are characterized by periods of eruption alternating with periods of dormancy of variable length, but there are many volcanoes that have a persistent constant level of activity (such as Etna and

Volcano alert level with associated aviation colour code showing increasing level of concern from green (normal) to red (warning), signalling an imminent or an underway eruption.

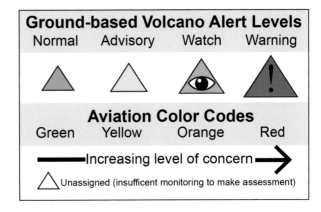

Stromboli in Italy; Popocatépetl and Colima, Mexico; and Hawaii, USA). The Global Volcanism Program of the Smithsonian Institution, National Museum of Natural History provides a list of currently active volcanoes and other useful information (https://volcano.si.edu/).

There are four main early warning signs of volcanic unrest: seismicity, ground deformation, degassing and thermal anomaly. In some cases, a fifth signal can be recorded in changes of the lava lake (if there is any), like the one that occurred in July 2018 at Halema`uma`u crater, the summit of Kilauea caldera. Monitoring a volcano involves measuring many different parameters that together can help to determine the status of the volcano.

Seismicity is one the main signs of volcanic unrest and monitoring volcanic earthquakes is a principal activity of every volcano observatory. Volcanic earthquakes are usually of lower intensity than tectonic earthquakes and they can go undetected if there is no monitoring network in place. A good seismic network at a volcano also allows volcanologists to distinguish between different type of seismic signals, which in turn can be attributed to specific volcanic events and volcanic processes. The movement of magma, fluids and gas generate particular seismic waves and so does the breaking of rock because of increasing stress at depth. Surface events such as rock falls, avalanches and lahars produce specific seismic signals, too, and they can also be detected. Therefore, a seismic network is fundamental for short-term volcanic hazards.

Pressure changes in the magma reservoir or in the overlying geothermal reservoir can cause ground deformation with uplift or subsidence. Periods of ground uplift can precede volcanic eruptions or can characterize periods of volcanic unrest without necessarily culminating in an eruption. Ground deformation episodes have been frequently observed at Campi Phlegrei caldera in southern Italy testified by the spectacular change in the ground level of the Solfatara in Pozzuoli, or at Santorini volcano, Greece in 2011. Ground deformation can be measured with ground tiltmeters, laser electronic distance measuring, high-precision levelling or via the Global Positioning System (GPS). Thanks to modern satellites volcanologists can monitor ground deformation at even the most remote volcanoes. In fact, there is less need for ground instrumentation, which was limiting our capacity to measure volcano deformation, since it was impossible to put ground instrumentation at every single volcano in the world. As a result of this increased capacity of monitoring the number of known deforming volcanoes (considered to be in a state of unrest), has increased fivefold in about 10 years.

Archaeological ruins of the ancient temple of Serapide, flooded by water, Pozzuoli, Naples, Italy. The level of flooding is changing with time due to ground deformation (uplift and subsidence) associated with the activity of Campi Flegrei caldera.

We are all familiar with fumaroles inside a volcano's crater or close to the summit. A fumarole is just an opening through which volcanoes emit gases. Indeed, as we have seen in chapter 2, volcanic gases are one of the fundamental components of magma and they play a big role in determining the style of a volcanic eruption. Therefore, it is not surprising that measurements of volcanic gases, their composition, their fluxes (i.e. the rate of gas flow and how it changes), their temperatures, and how these parameters change with time, are all very important in volcano monitoring. Measuring gas compositions, temperatures and fluxes has changed a lot in recent years. We have moved from in-situ measurements – requiring scientists to get very close to the fumaroles and spend some time there to do the measurements and take samples, putting themselves at high risk – to remote instruments that allow the volcanic gases and the volcanic plume to be measured from satellites and from a distance. Ground-based and in-situ measurements are still incredibly valuable, but remote techniques are becoming more routinely used.

Volcano monitoring is fundamental, particularly for short-term volcanic hazards. It provides a baseline of activity giving insights and patterns, which can be used to draw an emergency plan and to alert the population. However, with hundreds of volcanoes either active or in unrest, most of them completely unknown or in remote parts of the globe, it is simply impossible to monitor every single volcano. A recent study revealed that of 441 active volcanoes in 16 countries, about 384 have insufficient monitoring or no monitoring at all. As we have seen a period of unrest will not necessarily culminate in an eruption, and determining if a period of unrest will culminate in an eruption or not remains one of the main challenges for the scientific community and the volcano observatories.

A FORENSIC APPROACH TO VOLCANOLOGY

Volcano monitoring is fundamental for understanding what the volcano is doing right now, getting a baseline of 'normal' behaviour and identifying a period of unrest. However a good knowledge of the past eruptive behaviour of a volcano is fundamental to understand how the volcano works. This knowledge can then be used to interpret monitoring data, define possible scenarios for future eruptions with different styles and sizes and quantify uncertainty. It can also be used to draw maps showing different types of volcanic hazards. All this will contribute towards longer-term forecasts and therefore towards a more efficient management of the risk. Geological, petrological and geochronological data define what is now called a forensic approach to volcanology. Scientists looking at the rock record produced by a volcano learn from the past to understand the present and hopefully have glimpse of the future activity.

Reconstructing the geological record along with geochronological data (i.e. dating the rock record) allows us to draw up the stratigraphy of the volcano and get an idea of the past eruptive activity such as: the type of eruptions and frequency; the cycle of eruptive activity (i.e. if the volcano activity changes from effusive to explosive in a more or less cyclic pattern); the type of past volcanic hazards (i.e. lava flows, pyroclastic flows, ash fall, etc); the area affected by the different types of volcanic activity; the length of the eruptive activity and so on. Such data can be acquired for the entire volcanic history or in a more detailed way for a single eruption and are used to produce very useful maps, that range from a volcanic map (showing the entire eruptive history of the volcano) to specific hazard maps showing the dispersion area of ash or tephra fall, pyroclastic flows or lava flows during a specific eruption. They can then be used, sometimes combined with numerical simulation modelling, to produce maps of various volcanic hazards for use in emergency plans and also to inform the population living around the volcano.

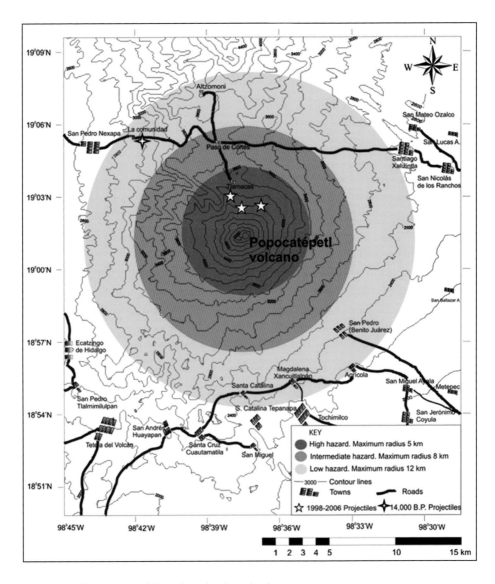

Hazard map of ballistic material (e.g. volcanic bombs and rocks fragments > 64 mm, 2½ in diameter) for Popocatépetl volcano, Mexico. The coloured area represents increasing probability of occurrence and therefore increasing risk from yellow to red. Locations with 5-point stars represent observed impact sites from 1998–2006; the 4-point star represents an impact site dating to 14,000 years ago.

MINERALS: AN ARCHIVE OF INFORMATION

Petrology is the study of the composition and genesis of rocks. Igneous petrologists study rocks formed from magma and therefore those formed by volcanoes. The composition of a volcanic rock can reveal a lot about the tectonic and magmatic setting of a volcano (see chapter 2, Volcanoes), but we can get even more information by looking at the minerals crystallized from the magma, the so-called 'crystal cargo'. Indeed, minerals crystallize from the magma stored underneath the volcano in the plumbing system. Most of the minerals forming a volcanic rock crystallize before the eruption and they are truly an archive of information. They are the volcano's messengers.

Retrieving the information locked in a crystal is not the easiest thing. Petrologists start by examining the rock under a petrographic microscope. To do this they use a petrographic thin section, which is a very thin slice of rock cut until it is transparent to light. By looking at the thin section, petrologists can recognize the different minerals that constitute the rock and retrieve other useful information

Volcanic rock under the petrographic microscope. The light minerals are called plagioclase, with chemical formula $(Na,Ca)(Al,Si)_4O_8$, the colored minerals are pyroxene $(Ca,Mg,Fe)_2Si_2O_6$.

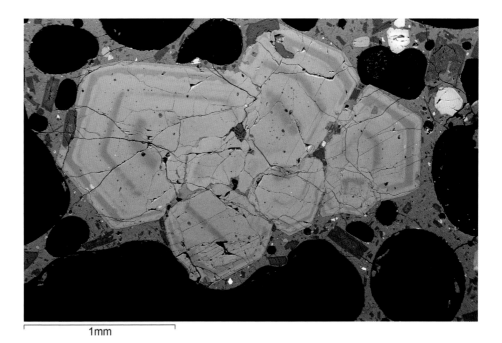

1mm

A volcanic rock from Stromboli volcano, Italy imaged under a high-
resolution electronic microscope. The group of grey minerals are pyroxene
and the different shades of grey indicate different portions of the crystal
with different composition. The minerals are surrounded by a fine grained
grey groundmass composed of tiny crystals and volcanic glass.

such as the texture of the rock and the relationship between minerals. For example, bigger minerals must have formed earlier than the smaller ones, as they had more time to form. From this the petrologist, for example, can work out an order of crystallization, which gives important information as to what happened before the eruption. We can also see how the crystallization sequence changed with the composition of the rock, so we can follow the evolution of the magma, which is the change in composition from less silica-rich basalt, to a more silica-rich magma while it is stored in magma reservoirs (see pp.38–39 on magma).

Modern analytical techniques can measure the chemical composition of a single mineral with very high precision and high spatial resolution. Most of the modern analytical techniques are non-destructive, which means that the mineral is not destroyed by the analysis and different techniques can be used to analyze the same spot. In this way we can unravel important information locked in the mineral structure and composition. We are able to reconstruct the physical conditions (pressure and temperature) and the chemical composition of the magma from which the crystal grew, we are also able to measure the volatile content of the magma before eruption, which is very important when dealing with explosive eruptions. In the last 20 years or so, an important development has enabled the retrieval, in some cases, of another important piece of information: the length of time that the mineral has spent in a particular magmatic environment before eruption.

Minerals grow from magma beneath the volcano before an eruption and, in a similar way to tree rings, they record the history of magma composition before the eruption. When new magma, which might have a different composition and different temperature, arrives in the magmatic reservoir and mixes with the resident magma, the minerals continue to grow but with a different composition to reflect the new magma. They form a compositional band, in a way similar to a new ring in a tree trunk. Analyzing the different compositional bands enables the reconstruction of the life-history of the mineral as it is subjected to different physico-chemical conditions, particularly temperature, for different lengths of time, before finally being ejected in an eruption. The information that we can gather in this way is very useful to understand how long it takes to build up an eruption. Having this information for as many as possible past eruptions of a volcano will give a better picture of how the volcano behaves and can be extremely useful when combined with monitoring data to inform the volcanic hazard assessment and forecast future eruptions.

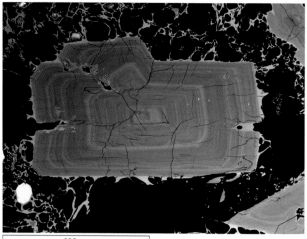

600μm

A crystal of compositionally zoned plagioclase surrounded by volcanic glass (grey filament) in a pumice from Popocatépetl volcano. The image is taken using a high-resolution electronic microscope and the different shades of grey indicate different portions of the crystal with different composition. The mineral grows from the core to the rim in a concentric way, changing composition in response to the environment, in a similar way to tree rings.

References

CHAPTER 2

Annen, C., Blundy, J.D. and Sparks, R.S.J., 2006. The genesis of intermediate and silicic magmas in deep crustal hot zones. *J. Petrol.*, 4:505–539.

Bachmann, O., Miller, C.F. and de Silva, S.L., 2007. The volcanic-plutonic connection as a stage for understanding crustal magmatism. *J. Volcanol. Geoth. Res.*, 167:1–23.

Campbell, I.H., 2005. Large Igneous Provinces and the mantle plume hypothesis. *Elements*, 1:265–269.

Cashman, K.V., Sparks, R.S.J. and Blundy J.D., 2017. Vertically extensive and unstable magmatic systems: a unified view of igneous processes. *Science*, 355:1280.

Coffin, M.F. and Eldholm, O., 1994. Large igneous provinces: crustal structure, dimensions, and external consequences. *Rev. Geophys.*, 32(1):1–36.

Enrst, R.E., Buchan, K.L. and Campbell, I.H., 2005. Frontiers in Large Igneous Province Research. *Lithos*, 79:271–297.

Heiken, G., 2013. *Dangerous Neighbors: Volcanoes and cities.* Cambridge University Press, 185pp.

Ingle, S. and Coffin, M.F., 2004. Impact origin for the greater Ontong Java Plateau? *Earth Planet Sc. Lett.*, 218:123–134.

Jerram, D., 2011. Introducing Volcanology: A Guide to Hot Rocks. *Dunedin*, 118pp.

Miller, C.F. and Wark, D.A., 2008. Supervolcanoes and their explosive supereruptions. *Elements*, 4:11–16.

Oppenheimer, C., 2001. Eruptions that Shook the World. *Cambridge University Press*, 406pp.

Jones, A.P., 2005. Meteorites impact as trigger to Large Igneous Province. *Elements*, 1:277–281.

Jones, G.S., Gregory, J.M., Stott, P.A., Tett, S.F. and Thorpe, R.B.,

2005. An AOGCM simulation of the climate response to a volcanic super-eruption. *Clim. Dynam.*, 25:725–738.

Parfit, E.A. and Wilson, L., 2009. *Fundamental of Physical Volcanology.* Blackwell Publishing, 230pp.

Rampino, M.R. and Self, S., 1992. Volcanic winter and accelerated glaciation following the Toba super-eruption. *Nature*, 359:50–52.

Robock, A., 2000. Volcanic eruptions and climate. *Rev. Geophys.*, 38:191–219.

Rogers, G.C., 1982. Oceanic plateaus as meteorite impact signatures. *Nature*, 299:341–34.

Robock, A., Ammann, C.M. and Oman, L., *et al.*, 2009. Did the Toba volcanic eruption of the ~ 74 ka B.P. produce widespread glaciations? *J. Geophys. Res.*, 114.

Rose, W.I. and Chesner, C.A., 1990. Worldwide dispersal of ash and gases from Earth's largest known eruptions: Toba, Sumatra, 75 kyr. *Paleogeog. Paleoclim. Paleoecol.*, 89:269–275.

Saunders, A.D., 2005. Large Igneous Province: origin and environmental consequences. *Elements*, 1:259–263.

Schmincke, H.U., 2004. *Volcanism.* Springer, 324pp.

Self, S., Thordason, T. and Widdowson, M., 2005. Gas fluxes from flood basalt eruptions. *Elements*, 1:283–287.

Sigurdsoon, H. (ed.), 2015. The Encyclopedia of Volcanoes (2ⁿᵈ ed.). *Academic Press*, 1421pp.

Sparks, R.S.J., Annen, C., Blundy, J.D., Cashman, K.V., Rust, A.C. and Jackson, M.D., 2019. Formation and dynamics of magma reservoirs. *Phil. Trans. R. Soc.* A 377:20180019.

Wignall, P., 2005. The link between Large Igneous Province and mass extinctions. *Elements*, 1:293–297.

CHAPTER 3

Giardini, D., Woessner, J., Danciu, L., Crowley, H., Cotton, F., Grünthal, G., Pinho, R. and Valensise, G., and the SHARE consortium. SHARE European Seismic Hazard Map for Peak Ground Acceleration, 10% Exceedance Probabilities in 50 years. https://www.researchgate.net/figure/The-SHARE-European-Seismic-Hazard-Map-displays-the-ground-motion-ie-the-Peak-Ground_fig4_314284870.

Mercalli, G., 1902. Sulle modificazioni proposte alla scala sismica De Rossi-Forel. *Boll. Soc. Sismol. Ital.*, 8, 184–191.

Musson, R., 2002. *Intensity and Intensity Scales, IASPEI New manual of seismological observatory practice*, chp 12, pp.653–672 (ed.) Bormann P., GeoForschungsZentrum Potsdam, Potsdam.

Reid, H.F., *The Mechanics of the Earthquake, The California Earthquake of April 18, 1906, Report of the State Investigation Commission* (vol 2). Carnegie Institution of Washington, pp.16–28.

Richter, C.F., 1935. An instrumental earthquake magnitude scale. *B. Seismol. Soc. Am.*, 25 (1–2):1–32.

Sieberg, A., 1923. *Geologische, Physikalische und Angewandte Erdbebenkunde.* G. Fischer, Jena.

Wood, H.O. and Neumann, F., 1931. *Modified Mercalli Intensity Scale of 1931. B. Seismol. Soc. Am.*, 21:277–283.

CHAPTER 4

Ambrose, S.H., 1998. Late Pleistocene human population bottlenecks, volcanic winter, and differentiation of modern humans. *J. Hum. Evol.*, 34:623–651.

Arndt, N.T., Fontboté, L., Hedenquist, J.W., Kesler, S.E., Thompson, J.F.H. and Wood,

D.G., 2017. Future Global Mineral Resources. *Geochem. Perspect.*, 6(1):171.

WKD Matsuo Basho Archive: https://matsuobasho-wkd. blogspot.com/2012/06/fuji-mount-fujisan.html.

BBC: https://www.bbc.co.uk/ programmes/b00llpvp.

Bonasia, R., Scaini, C., Capra, L., Nathenson, M., Siebe, C., Arana-Salina, L. and Folch, A., 2014. Long-renge hazard assessment of volcanic ash dispersal for a Plinian eruptive scenario at Popocatépetl volcano (Mexico): implications for civil aviation safety. *Bull. Volcanol.*, 76:789.

Calas, G., 2017. Mineral resources and sustainable development. *Elements*, 13:301–306.

Duffield, W.A., 2005. Volcanoes, geothermal energy, and the environment. In: Martí, J., and Ernst, G.G. (eds.), *Volcanoes and the Environment*. Cambridge University Press, chp 11, pp.304–332.

Endicott, P., Ho, S.Y.W., Metspalu, M. and Stringer, C., 2009. Evaluating the mitochondrial timescales of human evolution. *Trends Ecol. Evol.*, 24:515–521.

Espinasa-Pereña, R. and Martin-Del Pozzo, A.L., 2006. Morphostratigraphic evolution of Popocatépetl volcano, Mexico. *Geol. Soc. Am.*, Sp. Paper 402:101–123.

Gagneux P., Wills C., Gerloff U., *et al.* 1999. Mitochondrial sequences show diverse evolutionary histories of Africa hominoids. *Proc. Natl., Acad. Sci.*, 96:5077–5082.

General Kinematics: https://www.generalkinematics.com.

Gibson, H.L., 2005. Volcano-hosted ore deposits. In: Martí, J. and Ernst, G.G. (eds.), *Volcanoes and the Environment*. Cambridge University Press, chp 12, pp.333–386.

Global Volcanism Program of the Smithsonian Institution, National Museum of Natural History: http://volcano.si.edu.

Gorokhovich, Y., 2005. Abondonment of Minoan places on Crete in relation to the earthquake induced changes in gorudnwater supply. *J. Archol. Scie.*, 32:217–222.

King, G., Bailey, G. and Sturdy, D., 1994. Active tectonics and human survival strategies. *J. Geophys. Res.*, 99 (B10), 20,063–20,078.

Loughlin, S.C., Sparks, S. and Jenkins, S.F., 2015. *Global Volcanic Hazards and Risk*. Cambridge University Press, 409pp.

Louys, J., 2007. Limited effect of the Quaternary's largest super-eruption (Toba) on land mammals from Southeast Asia. *Quat. Sci. Rev.*, 26:3108–3117.

Luterbacher, J. and Pfister, C., 2015. The year without a summer. *Nat. Geosci.*, 8:246–248.

Martin, W., Baross, J., Kelley, D. and Russell, M., 2008. Hydrothermal vents and the origin of life. *Nat. Rev. Microbiol.*, 6:805–814.

McDougall, I., Brown, F.H. and Fleagle, J.C., 2005. Stratigraphic placement and age of modern humans from Kibish, Ethiopia. *Nature*, 433:733–736.

Oppenheimer, C., 2012. *Eruptions that Shook the World*. Cambridge University Press, 392pp.

Oppenheimer, C., 2002. Limited global change due to the largest known Quaternary eruption, Toba ~ 74 kyr BP? *Quatern. Scie. Rev.*, 21:1593–1609.

Orkustofnun National Energy Authority: https://nea.is/geothermal/.

Planket, P., Uruñuela, G., 2006. Social and cultural consequences of a late Holocene eruption of Popocatépetl in central Mexico. *Quat. Internal.*, 151:19–28.

Planket, P., Uruñuela, G., 2008. Mountain of sustenance, mountain of destruction: the prehispanic experience with Popocatépetl volcano. *J. Volcan. Geotherm. Res.*, 170:111–120.

Pyle, M.D., 2017. *Volcanoes: Encounters through the Ages*. Bodleian Library, 223pp.

Robock, A., 2000. Volcanic eruptions and climate. *Rev. Geophys.*, 38(2):191–219.

Robocl, A. and Mao, J., 1992. Winter warming from large volcanic eruptions. *Geophys. Res. Lett.*, 19(24): 2405–2408.

Scerri e.M.L., Thomas M., Manica A., Gunz P., *et al.*, 2018. Did our species evolve in subdivided populations across Africa, and why does it matter? *TREE*, 2399.

Self, S., 2005. Effects of volcanic eruptions on the atmosphere and climate. In: Martí, J. and Ernst, G.G. (eds.). *Volcanoes and the Environment*. Cambridge University Press, chp 5, pp.152–174.

Siebe, C., 2000. Age and archeological implications of Xitle volcano, southwestern Basin of Mexico City. *J. Volcan. Geotherm. Res.*, 104:45–64.

Siebe, C., Salinas, S., Arana-Salinas, L., Macias, J.L., Gardner, J. and Bonasia, R., 2017. The ~23,500 y ^{14}C BP White Pumice Plinian eruption and associated debris avalanche and Tochimilco lava flow of Popocatépetl volcano, Mexico. *J. Volcanol. Geotherm. Res.*

Smith, E.I., Jacobs, Z., Johnes, R. and Ren, M., *et al.*, 2018. Humans thrived in South Africa through the Toba eruption about 74,000 years ago. *Nature*, 555:511–515.

Stetter, O.S., 2005. Volcanoes, hydrothermal venting, and the origin of life. In: Martí, J. & Ernst, G.G. (eds.). Volcanoes and the Environment. Cambridge University Press, chp 6, pp.175–206.

USGS: https://www.usgs.gov.

Wheeler, D. and Demarée, G., 2005. The weather of the Waterloo campaign 16 to 18 June 1815: did it change the course of history? *Weather*, 60(6):159–164.

Williams, M., Ambrose, S.H., van der Kaars, S., Ruehlemann, C., Chattopadhyaya, U.C., Pal, J.N. and Chauhan, P., 2009. Environmental impact of the 73 ka Toba super-eruption in South Asia. *Palaeogeo. Palaeoclimatol. Palaeoecol.*, 284:295–314.

Winchester, S. 2003. *Krakatoa: The Day the World Exploded*. Penguin, 432pp.

Zielinski, G.A., Mayweski, P.A. and Meeker, I.D., *et al.*, 1996. Potential atmospheric impact of the Toba mega-eruption ~71,000 years ago. *Geophys. Res. Lett.*, 23(8):837–840.

CHAPTER 5

Allard, P., Baxter, P., Hallbwachs, M. and Komorowski, J., 2002. Nyiragongo. *Bull. Global Volcan. Network*, 27.

Biggs, J. and Pritchard, M.E., 2017. Global volcano monitoring: what does it mean when volcanoes deform? *Elements*, 13(1):17–22.

Brown, S.K., Loughlin, S.C., Sparks, R.S.J., Vye-Brown, C., Barclay, J., Calder, E., Cottrell, E., Jolly, G., Komorowski, J.C., Mandeville, C., Newhall, C., Palma, J., Potter, S. and Valentine, G., 2015. Global volcanic hazard and risk. In: Loughlin, S.C., Sparks, S. and Jenkins, S.F. *Global Volcanic Hazards and Risk.* Cambridge University Press, chp 2, pp.81–172.

Carn, S.A., Pallister, J.S., Lara, L., Ewert, J.W., Watt, S., Prata, J., Thomas, R.J. and Villarosa, G., 2009. The unexpected awakening of Chaitén Volcano, Chile. *EOS Transaction Am. Geophys. Union.*, 90(24):205–212.

Cooper, K.M., 2017. What does a magma reservoir look like? The 'crystal's eye' view. *Elements*, 13(1):23–28.

Fearnley, C.J., Birds, D.K., Haynes, K., McGuire, W.J. and Jolly J. (eds.), 2018. Observing the Volcano World. Springers Open.

Gottsmann, J., 2015. Volcanic unrest and short-term forecasting capacity. In: Loughlin, S.C., Sparks, S. and Jenkins, S.F. *Global Volcanic Hazards and Risk.* Cambridge University Press, chp 18, pp.317–321.

Harris, A.J.L., 2015. Basaltic lava flow hazard. In: Papale, P. (ed.). *Volcanic Hazards, Risks and Disasters.* Elsevier Hazards and Disaster Series, chp 2, pp.17–46.

Heiken, G., 2013. *Dangerous Neighbors Volcanoes and Cities.* Cambridge University Press, 185pp.

Komorowski, J.C. and Karume, K., 2015. Nyiragongo (Democratic Republic of Congo), January 2002: a major eruption in the midst of a complex humanitarian emergency. In: Loughlin, S.C., Sparks S. and Jenkins, S.F. *Global Volcanic Hazards and Risk.* Cambridge University Press, chp 11, pp.273–280.

Jenkins, S.F., Wilson, T., Magill, C., Miller, V., Stewart, C., Blong, R., Marzocchi, W., Boulton, M., Bonadonna, C. and Costa, A., 2015. Volcanic ash fall hazard and risk. In: Loughlin, S.C., Sparks, S. and Jenkins, S.F. *Global Volcanic Hazards and Risk.* Cambridge University Press, chp 3, pp.173–221.

Lara, L.E., 2009. The 2008 eruption of the Chaitén Volcano, Chile: a preliminary report. *Andean Geology*, 36(1):125–129.

Matrolorenzo, G., Petroni, P.P., Pagano, M., Incoronato, A., Baxter, P.J., Canzanella, A. and Fattore, L., 2001. Herculaneum victims of Vesuvius in AD 79. *Nature*, 410:769–770.

Neri, A., Esposti Ongaro, T., Voight, B. and Widiwijayanti, C., 2015. *Pyroclastic Density Current Hazards and Risk.* Elsevier Hazards and Disaster Series, chp 5, pp.109–140.

Newhall, C. and Hoblitt, R., 2002. Constructing event trees for volcanic crises. *Bull. Volcanol.*, 61(1):3–10.

Osservatorio Vesuviano: http://www.ov.ingv.it/ov/en.html.

Parks, M.M., Biggs J. and England, P., *et al.*, 2012. Evolution of Santorini volcano dominated by episodic and rapid fluxes of melt form depth. *Nat. Geosci.*

Philipson, G., Sobradelo and R., Gottsmann, J., 2013. Global volcanic unrest in the 21st century: an analysis of the first decade. *J. Volcanol. Goetherm Res.*, 264:183–196.

Petrone, C.M., 2018. Volcanic eruptions: From ionosphere to the plumbing system. Geology, 46:927–928.

Petrone, C.M., Bugatti, G., Braschi, E. and Tommasini, S., 2016.

Pre-eruptive magmatic processes re-timed using a non-isothermal approach to magma chamber dynamics. *Nat. Commun.* 7:12946.

Petrone, C.M., Braschi, E., Francalanci, L., Casalini, M. and Tommasini S., 2018. Rapid mixing and short storage timescales in the magma dynamics of a steady-state volcano. *EPSL*, 492:206–221.

Pompeii: *http://www.pompeii.org.uk.*

Saccorotti, G., Iguchi, M. and Aiuppa, A., 2015. *In situ Volcano Monitoring: Present and Future.* Elsevier Hazards and Disaster Series, chp 6, pp.141–168.

Sigurdsson, H., Houghton, B., McNutt, S.R., Rymer, H. and Stix, J., 2015. The Encyclopedia of Volcanoes (2nd ed.). *Elsevier*, 1421pp.

Sparks, R.S.J., Biggs, J. and Neuberg, J.W., 2012. Monitoring Volcanoes. *Science*, 335:1310–1311.

Tanguy, J.C., 1994. The 1902–1905 eruptions of Mt Pelée, Martinique: anatomy and retrospection. *J. Volcanol. Geotherm. Res.*, 60(2):87–107.

Tedesco, D., Vaselli, O., Papale, P., Carn, S., Voltaggio, M., Sawyer, G., Durieux, J., Kasereka, M. and Tassi, F., 2007. January 2002 volcano-tectonic eruption of Nyiragongo volcano, Democratic Republic of Congo. *J. Geophys. Res.: Solid Earth* (1978–2012), 112.

USGS HVO: https://volcanoes.usgs.gov/observatories/hvo/.

USGS Volcanic Ash: https://volcanoes.usgs.gov/volcanic_ash/.

USGS Volcanoes: https://www.usgs.gov/news/k-lauea-volcano-erupts.

Wilson, T.M., Jenkins, S. and Stewart, C., 2015. Impacts from volcanic ash fall. In: Papale, P. (ed.). *Volcanic Hazards, Risks and Disasters.* Elsevier Hazards and Disaster Series, chp 3, pp.47–86.

Zanchetta, G., Sulpizio, R., Pareschi, M.T., Leaoni, F.M. and Santacroce, R., 2004. Characteristics of May 5–6, 1998 volcaniclastic debris flows in the Sarno area (Campania, Southern Italy): relationship to structural damage and hazard zonation. *J. Volcanol. Geotherm. Res.*, 133:377–393.

Index

Acknowledgments

Writing this book has been fun and adventurous, with some twists and turns and new companions found on the way. I would like to thank my co-authors Roberto Scandone and Alex Whittaker for their amazing contributions. I also would like to thank Richard Herrington for encouraging me to write this book. I am very grateful to Hilary Downes and Chris Stanley, who helped to improve the text. I am indebted to Daniele Andronico and Lisetta Giacomelli who have generously shared some of their amazing photos of volcanoes from around the world, and Eleonora Braschi who helped with a desperate search for a photo. The moral support of Sara Russell and Paul Schofield has been of great help. The skilful editorial handling and understanding by Trudy Brannan and her team is greatly appreciated. I am also indebted to my colleagues that have inspired me with their work and their presence. There are too many to mention here, but I hope you know who you are. Finally, I would like to thank my partner Betty for her constant support and encouragement, and my parents for being there even when it is difficult.

CHIARA MARIA PETRONE

Picture credits

p.5 ©Daniele Andronico; p.7, 35, 65, 75, 120, 124, 132 ©Photo courtesy of L. Giacomelli and R. Scandone; p.10 Adapted from Wadsworth Publishing Company/ITP; p.14, 15, 21, 49 Adapted from USGS; p.17 Adapted from Bates, 1980; p.19 (top) Adapted from *The Sciences*, 6th ed., by Trefil and Hazen; p.19 (bottom), 23 (top) ©Claus Lunau/Science Photo Library; p.23(bottom) ©Gary Hicks/Science Photo Library; p.25 ©Muller, R.D., M. Sdrolias, C. Gaina, and W.R. Roest, 2008, Age, spreading rates and spreading symmetry of the world's ocean crust, *Geochem. Geophys. Geosyst.*, 9, Q04006, doi:10.1029/2007GC001743. [CC BY-SA 3.0 (http://creativecommons.org/licenses/by-sa/3.0/)]; p.30 ©mTaira/Shutterstock.com; p.31 ©Ruslan Shaforostov/Shutterstock.com; p.32 ©U.S. Navy photo by Philip A. McDaniel; p.34 ©Wead/Shutterstock.com; p.37 © William Crochot [CC BY-SA 4.0 (https://creativecommons.org/licenses/by-sa/4.0/)]; p.39 (top) Eleonora Braschi; p.39 (bottom) © Gyvafoto/Shutterstock.com; p.42(left) ©Caricchi L, Blundy J, The temporal evolution of chemical and physical properties of magmatic systems, Geological Society, London, Special Publications, 422, 1-15, 19 August 2015; p.43 ©bierchen/Shutterstock.com; p.44 © Nathan Mortimer/Shutterstock.com; p.45 ©orxy/Shutterstock.com; p.46 (top) ©Terence Mendoza/ Shutterstock.com; p.46 (bottom) ©Omdrej Prochazka/Shutterstock.com; p.47 ©Nick Upton.naturepl.com; p.48 ©Attila JANDI/Shutterstock.com; p.51 Adapted from Cas and Wright, 1988; p.52, 53(bottom), 56, 57(bottom), 59(bottom), 60(bottom), 62, 63, 107, 108, 127(bottom), 135, 136, 137 ©Chiara Maria Petrone; p.53 (top) Wellcome Collection; p.54 ©balounm/Shutterstock.com; p.58 Adapted from Cross-sectional model of a Plinian Eruption, generated by Jeff Sale and Vic Camp; p.59(top) ©udaix/Shutterstock.com; p.60(top) ©Pete Oxford/naturepl.com; p.61 ©Jeff Zenner Photography/Shutterstock.com; p.67 Adapted from Christyyc/Wikimedia Commons [CC BY-SA 4.0 (https://creativecommons.org/licenses/by-sa/4.0/)]; p.70 Adapted from Courtillot and Renne 2003; p.72 ©US Geological Survey/Science Photo Library; p.74 ©Dr Morley Read/Shutterstock.com; p.77 ©Lawrence Braile; p.78 ©James King-Holmes/Science Photo Library; p.81 Adapted from

Preliminary Reference Earth Model of Dziewonski and Anderson 12; p.86, 89, 116, 123, 130 ©U.S. Geological Survey; p.87 Ungtss at English Wikipedia. [Public domain]; p.88 ©2013, ETH Zurich on behalf of the EU-FP7 Consortium of EFEHR; p.91 ©Digital Globe/Science Photo Library; p.92 ©USGS/Cascades Volcano Observatory; p.95 Adapted from top image p.154, J. Marti and G.J. Ernst, *Volcanoes and their Environment*, Cambridge University Press, 2005; p.97 (top) ESA; p.97 (bottom) RethaAretha/shutterstock.com; p.100 (top) Adapted from National Energy Authority of Iceland; (bottom) Thermal Infrared Remote Sensing of Geothermal Systems Christian Haselwimmer and Anupma Prakash; p.103 ©R.M. Nunes/Shutterstock.com; p.106 © Siebe et al., 2017, The ~ 23,500 y 14C BP White Pumice Plinian eruption and associated debris avalanche and Tochimilco lava flow of Popocatépetl volcano, México, *J. Volcanol. Geothermal Res.*, Vols 333–334, p.66-95; p.109 ©Plunket, P., Urunuela, G., 2006, Social and cultural consequences of a late Holocene eruption of Popocatépetl in central Mexico, *Quater. Internat.*, Vol 151, Issue 1, p19-28; p.110 Katsushika Hokusai [Public domain]; p.113 ©Aldivo Ahmad/Shutterstock.com; p.115 ©K. Narloch-Liberra/Shutterstock.com; p.118 ©Fredy Thurerig/Shutterstock.com; p.119 ©Marian Galovic/Shutterstock.com; p.122 ©Martin Rietze/Science Photo Library; p.125 ©birdymeo/Shutterstock.com; p.127 (top) ©tupatu76/Shutterstock.com; p.128 © Zanchetta et al., 2004, Characteristics of May 5–6, 1998 volcaniclastic debris flows in the Sarno area (Campania, southern Italy): relationships to structural damage and hazard zonation, *J. Volcanol. Geothermal Res.*, Vol 133, Issues 1-4, p377-393; p.134 © Alatorre-Ibargüengoitia, M.A., Delgado-Granados, H. & Dingwell, D.B. *Bull Volcanol.* (2012) 74: 2155.

Unless otherwise stated images copyright of Natural History Museum, London.

Every effort has been made to contact and accurately credit all copyright holders. If we have been unsuccessful, we apologise and welcome correction for future editions and reprints.